T0234576

Blind Image Deconvolution

Subhasis Chaudhuri • Rajbabu Velmurugan
Renu Rameshan

Blind Image Deconvolution

Methods and Convergence

 Springer

Subhasis Chaudhuri
IIT Bombay, Mumbai
Maharashtra, India

Rajbabu Velmurugan
IIT Bombay, Mumbai
Maharashtra, India

Renu Rameshan
IIT Bombay, Mumbai
Maharashtra, India

ISBN 978-3-319-35216-9   ISBN 978-3-319-10485-0 (eBook)
DOI 10.1007/978-3-319-10485-0
Springer Cham Heidelberg New York Dordrecht London

Printed on acid-free paper

Springer is part of Springer Science+Business Media (www.springer.com)

To

*Two very special renunciants – Kalipada Maharaj
and Swami Shantananda.
SC.*

*Karthi chitappa.
VRB.*

*All from whom I have learned.
RMR.*

Preface

The issue of blur in images has kept researchers in the area of image processing busy for over half a century, and the researchers at the Vision and Image Processing Laboratory, IIT Bombay, were no exceptions. We have been working in this area for the last two decades. Earlier our efforts were confined to investigating the beneficial aspects of blur in images wherein we had used defocus blur to estimate depth in a scene. A monograph entitled *Depth from Defocus: A Real Aperture Imaging Approach*, published by the same publisher in 1999, dealt with the problem of deblurring when there are multiple observations. We derived performance bounds as a function of relative blurring among multiple observations. Subsequently, in the monograph entitled, *Motion-Free Super-Resolution*, again published by Springer in 2005, we demonstrated how the relative blurring among various observations could be used to super-resolve an image very efficiently.

In this monograph we investigate the deblurring problem for a single observation. This problem is well known in literature as blind deconvolution, and there has been a large volume of work by various researchers as regards how one can perform deconvolution when the blur is not known. So the question does arise what purpose then does this monograph serve? Is this yet another book on image deblurring? Publication of this monograph does require a justification. A review of the published literature on blind deconvolution suggests that these works concentrate mostly on accuracy and efficiency of various proposed algorithms.

Since blind deconvolution has both the point spread function (PSF) and the image unknown, both of them are required to be estimated. Very often this requires solving an alternate minimization problem. However, solving an alternate minimization problem does not always lead to useful solutions. Sometimes it leads to trivial solutions. It is no wonder that a method proposed by one group does not work for another group as it requires several parameters to be tweaked. Hence there is a genuine need for analyzing convergence properties of alternate minimization methods. We perform precisely the same in this monograph so that the end user can pre-judge the applicability of a given blind deconvolution algorithm for a specific problem. This also helps in understanding the behavior of an algorithm as iterations proceed. We hope that the readers will find the analysis useful.

In order to make the monograph useful to the practitioners in the industry, we also investigate one of the current techniques in detail for blind deconvolution. This involves usage of sparsity as a constraint as opposed to typical image PSF priors such as Gibbs distribution. It can be seen that sparsity is, indeed, a good candidate method for blind deconvolution.

The intended primary audience of this monograph are the graduate students in mathematics, EE, and CS departments. This should also serve as a good reference book for researchers in the area of image processing. The practitioners in image processing area would also find the monograph useful as it will provide them with a good theoretical insight to the problem. We have tried to make the book self-contained and hence there is no pre-requisite. However, familiarity with basics of signal and image processing and linear algebra would be helpful to the readers.

We sincerely hope that the readers will find the monograph useful and we welcome comments and suggestions from the readers.

Mumbai, India   Subhasis Chaudhuri
April 2014   Rajbabu Velmurugan
                                                                               Renu Rameshan

Acknowledgements

The first author acknowledges the support provided by DST (India) in the form of a J. C. Bose Fellowship. The second and third authors would like to thank DST for financial support in the form of an Indo-Swiss collaborative project. The authors are also thankful to Prof. Sibi Raj B Pillai and Prof. B. Yegnanarayna for their comments and suggestions. The patience and support of the publisher are also gratefully acknowledged.

Finally all good things start at the family. Hence the authors are thankful to their family members for their support and understanding.

Mumbai, India
April 2014

Subhasis Chaudhuri
Rajbabu Velmurugan
Renu Rameshan

Contents

List of Symbols

$\| \cdot \|_2, \ell_2$	L2 norm
$\| \cdot \|_1, \ell_1$	L1 norm
$\otimes$	Kronecker product
y, $\underline{y}$	Observed blurred and noisy image (matrix and vector form, respectively)
x, $\underline{x}$	Original sharp image (matrix and vector form, respectively)
k	Point spread function
$\mathcal{N}$, $\underline{n}$	White Gaussian noise
$\circledast$	Convolution operation
K	Convolution matrix
$R_x(.)$	Image regularizer
λ_x	Image regularization factor
$R_k(.)$	PSF regularizer
λ_k	PSF regularization factor
η_{min}	Minimum eigenvalue of sum of convolution matrices arising from PSF and regularizer
γ_{min}	Minimum eigenvalue of sum of convolution matrices arising from image and regularizer
$d(.,.)$	non-negative function used to define three- and four-point properties
$g_{h,i}$, $g_{v,i}$	First order horizontal and vertical difference at pixel i
D	First order difference matrix
y'	First order difference of y

Chapter 1
Introduction

With the proliferation of digital cameras, amateur photography is in the rise and among the vast amount of digital images generated many are blurred due to camera shake. This is an expected phenomenon since light weight cameras are more prone to movement and unless a tripod is used the chances for blurring is high. Object motion and camera defocus can also lead to a blurred image. Similar scenarios arise in medical, biological and astronomical imaging. Blurring in medical imaging can occur due to the image recording process, due to the non-idealities in radiation propagation through the subject, and due to subject motion. Biological and astronomical imaging systems are diffraction-limited systems. Hence the image of a point source appears as a disc, and overlap of such discs from neighbouring pixels leads to blurring. In addition in systems like the wide-field microscope the light scattered from out of focus planes causes blurring and the out-of-focus light depends on the object being imaged. If the nature of the blur is known it is possible to use deconvolution to estimate the sharp image. In all the examples mentioned above it is difficult or impossible to model the blur. In such cases the only solution to obtain a sharper image is by using blind deconvolution, wherein one tries to reconstruct the original image without any knowledge of the way the camera/subject moved. In this monograph we focus on regularization based methods of blind deconvolution and their convergence analysis.

1.1 Image Degradation

During the process of image capturing two sources of degradation can arise – noise and blurring. Blurring occurs during the image capturing process and leads to spreading of edges causing loss of sharpness and contrast in the recorded image. The blurring process can be seen as low pass filtering. Noise occurs mostly during image recording at the charge gathering step in the CCD and due to quantization.

© Springer International Publishing Switzerland 2014
S. Chaudhuri et al., *Blind Image Deconvolution: Methods and Convergence*,
DOI 10.1007/978-3-319-10485-0__1

Blurring in an image can occur either due to the imaging system itself or due to relative motion between the scene and the camera. Ideally a point source should be imaged as a point in the image plane, but due to aberrations of the lens [146] or due to inherent diffraction effects, the point gets spread out. Even if the lens is perfect with no aberrations, a point source gets imaged as a circular disk because of diffraction at the aperture of the imaging system. The blur function or the impulse response of the imaging system is also called as the point spread function (PSF). Diffraction limited imaging systems are high resolution imaging devices where diffraction creates a substantial amount of blurring. Microscopes, telescopes, confocal scanning laser microscope (CSLM), etc., are examples of diffraction limited imaging systems. CSLM can do optical sectioning of a fluorescent object making three-dimensional imaging possible. In this case in addition to the aperture effects, the light scattered from the planes adjacent to the one that is imaged also adds to the blurring.

Another source of blurring is the defocus blurring, arising due to the finite depth-of-field of the imaging systems. The geometry of image formation process is shown in Fig. 1.1. For a thin lens, the focal length (f), object and image distances (μ, and v, respectively) from the lens satisfy the lens equation

$$\frac{1}{\mu} + \frac{1}{v} = \frac{1}{f}. \tag{1.1}$$

If the image plane is at a distance v, then each point in the object is mapped to a point in the image plane. When the image plane is at v_0 (Fig. 1.1), where the lens equation is not satisfied, a point in the object is mapped to a circular patch assuming a circular aperture. This leads to overlap of points leading to blurring. It was shown in [120] that the blur thus introduced can be modeled as a circularly symmetric 2-D Gaussian function

$$k(u, v) = \frac{1}{2\pi\sigma^2} \exp \frac{-(u^2 + v^2)}{2\sigma^2}, \tag{1.2}$$

where σ, the blur parameter depends on the scene depth, and (u, v) represents a point in space. While imaging a 3D scene, defocus blur is always a spatially varying blur, since different scene points have different depths leading to a blur which varies spatially.

In an imaging system, the image is formed by averaging the photon count over the time the shutter is open. This temporal averaging leads to motion blur when there is a relative motion between the scene imaged and the camera during exposure. In linear motion blur, the camera has only translational motion. For horizontal motion with uniform velocity the PSF is [58]

$$k(u, v) = \frac{1}{\alpha_0} \text{rect}\left(\frac{u}{\alpha_0} - \frac{1}{2}\right)\delta(v), \tag{1.3}$$

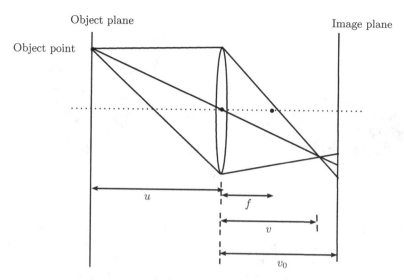

Fig. 1.1 Image formation showing defocus blurring. When the image plane is at v_0 the point light source gets imaged as a disk

where α_0 is the distance by which the camera moved while the shutter was open. Motion blur can also be rotational, or any arbitrary combination of translation and rotation, leading to images that are "shaken". In such cases modeling the PSF is a difficult task leaving one only with the option of doing deblurring without knowledge of PSF. Images degraded by a linear motion blur and rotational motion blur, which were captured by moving the camera are shown in Fig. 1.2a, b, respectively. The first image shows the effect of horizontal motion blur where the PSF can be defined by Eq. (1.3). The second image is blurred due to camera rotation and it cannot be represented by Eq. (1.3). This is an example of space varying blur and we do not address this problem in this monograph. In astronomical imaging, atmospheric turbulence is a source of blurring. This arises due to the random variation of refractive index in the medium between the object being imaged and the camera. The blur due to atmospheric turbulence can be modeled as [58]

$$k(u, v) = \exp(-\pi \alpha^2 (u^2 + v^2)), \tag{1.4}$$

where α determines the severity of the blur [123]. Noise is the other source of degradation in an image. Noise that occurs in a digital image can be either additive or multiplicative, signal-dependent or independent in nature. Signal-dependent noise is due to the random fluctuations in the number of photons detected at a point [4, 58]. The randomness arises since the photons arrive not as a steady stream but obey Poisson law and its effect is more visible under low lighting conditions. In this case each pixel can be modeled as a Poisson random variable with mean being the intensity of the pixel. For high intensities the Poisson distribution can be

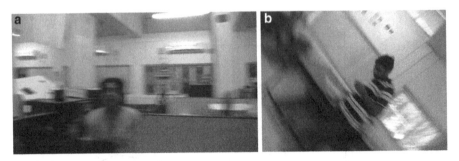

Fig. 1.2 Illustration of blur due to linear and rotational motion. (**a**) Linear motion blur. (**b**) Rotational motion blur

approximated by a Gaussian distribution. The additive component of noise comes from thermal noise in the sensor and the quantization noise, which can be modeled as white Gaussian noise. Another type of noise observed in coherent laser imaging is the speckle noise, which is mostly multiplicative in nature and occurs when the roughness of the surface being imaged is of the order of the wavelength of the incident radiation [4, 44]. We assume only additive white Gaussian noise (AWGN) for all the analysis in this monograph.

In this work we focus on restoration of images degraded by Gaussian and linear motion blur and containing AWGN. Having seen various models for blur, we next look at the image formation model.

1.2 Image Formation Model

The image intensity at a point (u, v), $u, v \in \mathbb{R}$ in the image due to the radiance at a location (u_1, v_1) depends on the image radiance at the neighborhood of (u_1, v_1). Assuming a linear image formation model, the intensity $y(u, v)$ is obtained as [129]

$$y(u, v) = \int_{-\infty}^{\infty} \int_{-\infty}^{\infty} k(u, v; u_1, v_1) x(u_1, v_1) du_1 dv_1, \qquad (1.5)$$

where $k(u, v; u_1, v_1)$ is the response at (u, v) of the image formation system to an impulse located at (u_1, v_1). $k(.)$ is called as the point spread function (PSF) of the system. If shift invariance is assumed, Eq. (1.5) gets modified as

$$y(u, v) = \int_{-\infty}^{\infty} \int_{-\infty}^{\infty} k(u - u_1, v - v_1) x(u_1, v_1) du_1 dv_1. \qquad (1.6)$$

Since the analysis and implementations are done in the discrete domain, Eq. (1.6) is discretized [129] to obtain

$$y(m,n) = \sum_{m'} \sum_{n'} k(m-m', n-n') x(m',n') + \mathcal{N}(m,n), \qquad (1.7)$$

where $y(m,n)$, $x(m,n)$, and $k(m,n)$ are the discrete forms of the observed image, original image and the PSF, respectively. $\mathcal{N}(m,n)$ is the AWGN which is inherent in the image recording system. This model for the imaging system is shown in Fig. 1.3. In this model the blurring process is assumed to be a linear system.

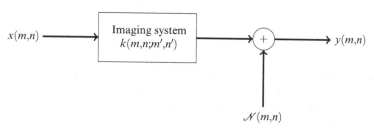

Fig. 1.3 Image formation model

The discrete quantities, namely the image and the PSF are represented as matrices with a size of $M \times N$ and $P \times Q$, respectively. It is assumed that PSF is of much smaller size than the image. The elements of k are positive and real and it is also assumed that the mean of the image does not change with convolution. These two conditions give the following constraints on the PSF matrix:

$$k(m,n) \geq 0, \qquad 0 \leq m \leq P, \ 0 \leq n \leq Q;$$
$$\sum_{m} \sum_{n} k(m,n) = 1. \qquad (1.8)$$

Written as a matrix-vector product Eq. (1.7) becomes

$$\underline{y} = K\underline{x} + \underline{n}, \qquad (1.9)$$

where K is the convolution matrix which is block circulant with circulant blocks (BCCB) when the convolution in Eq. (1.7) is circular. Here $\underline{x}$ and $\underline{y}$ are the vector forms of the image obtained by lexicographically ordering x and y, respectively.

1.3 Blind Deconvolution

In all the types of blurring described in Sect. 1.1, either the PSF is not known or even if the form of the PSF is known, its parameters are not. Hence the reconstruction has to proceed without the knowledge or with a partial knowledge of the PSF. To restate, the objective of blind deconvolution is to estimate the sharp image x and the PSF k, given the noisy blurred observation y and the noise statistics.

The image formation model in Eq. (1.9) enables us to calculate the image formed given the scene and the PSF. This calculation is a straight forward convolution operation which does not pose any difficulties. Since the image formation process is a low pass filtering operation the recorded image has lost some information which is present in the original scene. Blind deconvolution is an inverse problem corresponding to the direct problem of convolution. Let $\mathbf{X}$, and $\mathbf{Y}$ be Hilbert spaces and $\mathscr{K} : \mathbf{X} \to \mathbf{Y}$ be a transformation from $\mathbf{X}$ to $\mathbf{Y}$. Given

$$\mathscr{K}(x) = y, \ x \in \mathbf{X}, \ y \in \mathbf{Y}, \tag{1.10}$$

the forward problem is finding y from x and K. The inverse problem corresponds to estimating x, given y and K. In blind deconvolution, additionally K is unknown.

Inverse problems are inherently difficult to solve [9], owing to the fact that there is a loss of information as in the case when K is equivalent to low pass filtering where the high frequency information is lost. Inverse problems are typically ill-posed and this is true for blind deconvolution also. Any problem which is not well-posed is termed as an ill-posed problem. A problem is well-posed in the Hadamard sense [108] if the following three conditions are satisfied:

- A solution exists.
- The solution is unique.
- The solution is stable under perturbation of data, i.e., the solution depends continuously on data, or an infinitesimal change in the data should produce only a similar change in the solution.

For the inverse problem in Eq. (1.10) the first condition is equivalent to the forward transformation $\mathscr{K}$ being surjective. The condition of unique solution maps to $\mathscr{K}$ being injective. First and second conditions together imply that the inverse exists and the third condition requires the inverse to be a continuous function.

Of the three conditions, violation of the third leads to numerical instabilities. There are no mathematical methods to convert an inherently unstable problem to a stable one. The best that can be done is to solve an approximation to the problem by using regularizers. Regularization methods recover partial information about the solution as stably as possible [33]. Violation of first and second conditions does not pose serious problems as the violation of the third [33]. We investigate the reasons for ill-posedness of the blind deconvolution problem and methods to handle it in Chap. 2, followed by a survey of the existing literature. A detailed road map of the book is provided in the next section.

1.4 Organization of the Book

The previous sections show that in many practical scenarios the only solution to obtain a sharp image is by use of blind deconvolution. Blind deconvolution being an ill-posed problem, as mentioned by Engl et al. [33], "no mathematical trickery can make an inherently unstable problem stable". The best that one can do is using regularization methods to get approximate solutions.

Blind deconvolution is a well researched area and over the past 30 years significant contributions have been made towards tackling this problem. The mathematical concepts used in the monograph are discussed in Chap. 2. In this chapter we discuss the reason behind ill-posedness of the blind deconvolution problem and also how to handle ill-posedness using regularization techniques. In addition the statistical estimation methods such as maximum likelihood, maximum a posteriori probability estimation (MAP), and optimization techniques are also briefly discussed. Chapter 3 discusses in detail the key solutions proposed in the literature so far. Starting from the early spectral domain methods which focused on finding inverse filters, the discussion proceeds to statistical methods and regularization methods. Since a large volume of research done in this area focuses on blind motion deblurring we devote a section to it, discussing the recent advances. Even though a significant amount of work has been done in this area, algorithms or implementations which can efficiently deconvolve a real life image of large size are still lacking. In the following chapters we derive the conditions for selecting appropriate image and PSF regularizers and also analyze the convergence of alternate minimization based blind deconvolution algorithms.

Following the work by Fergus [36] and the recent paper by Levin et al. [2], there is a common belief that joint MAP for blind deconvolution leads to trivial results. In Chap. 4 we prove the suitability of joint MAP for blind deconvolution. It is true that the conventional gradient based priors for image favor the blurred image as the solution, when the estimated PSF is the discrete impulse (trivial solution). But this would happen only if an *appropriate* prior is not chosen for the PSF. An *appropriate* PSF prior is one which has a cost that attains its maximum value when the PSF is the discrete impulse. We show that if an appropriate prior is chosen for the PSF, the discrete impulse solution for PSF is ruled out. We prove this theoretically for a general gradient based regularizer for the PSF and as a specific case, for the total variation (TV) regularizer. In addition to choosing an appropriate prior, it should also be ensured that the regularization factor for PSF is appropriately chosen. We give a theoretical lower bound for the PSF regularization factor and an empirical upper bound. For values of regularization parameter below the lower bound, the solution converges to the trivial one and for values above the upper bound the PSF tends to the averaging filter. In the acceptable range of values for the regularization factor, the amount of deblurring does vary and we choose the best regularization factor experimentally.

Having shown that joint MAP does work for blind deconvolution, we proceed by selecting TV based priors for both the image as well as the PSF. Since both the image and the PSF are unknowns alternate minimization (AM) is used to solve the blind deconvolution problem. In Chap. 5 we provide a Fourier domain convergence analysis of the AM procedure. TV prior being non-linear, a non-quadratic cost function is obtained at each iteration of the AM. This makes the convergence analysis difficult. For making the analysis feasible, the TV prior is replaced by its quadratic upper bound [40]. With this approximation, the cost function at each step of AM becomes quadratic in nature. Even with this approximation, the system obtained is not a linear shift invariant (LSI) system, which makes analysis in the spectral domain difficult. To overcome this, we use a further approximation, which makes the system LSI in each iteration – with the system changing with iteration. It is observed that the resulting regularization is adaptive in nature. We also note that the resulting system behaves like an adaptive Wiener filter. Once the fixed-point is reached the regularization factors remain constant, and a Fourier domain analysis shows that the fixed-points for image and PSF magnitude are similar to that in [17]. Our analysis differs in that the image and the PSF regularization factors becomes signal (image/PSF) dependent. It is observed that the convergence point of the magnitude of the image and the PSF are related to each other, which is due to the approximation to make the system LSI. We analyze the error term in the approximation, which gives insight into the regularizing capability of the TV regularizer.

Fourier domain analysis is feasible only for regularizers which are quadratic in nature. Hence we consider a spatial domain convergence analysis in Chap. 6. Tusnady et al. [27] has proved the convergence of AM class of algorithms when the cost function is KL divergence [77] for problems that arise in information theory. A simplified version of Tusnady's work was used for proof of convergence of algorithms which were recast in an AM framework by Byrne [19]. The problems which were dealt with, in Byrne's work were all solvable without using the AM concept, but recast in the AM framework for demonstration of the three-point, four-point proof for AM convergence which was introduced in [19]. We use the three-point, four-point concept for proving that the AM algorithm for blind deconvolution converges to the infimum of the cost function. The analysis proceeds by looking at the reduction in the cost function when one variable is kept constant and the other is minimized. We show that the three-point property is satisfied for all points in the image space and the non-negative function needed for the definition of three- and four-point properties is derived. It is further proved that with proper choice of image and PSF step sizes and in the presence of sufficient regularization, the four-point property is satisfied for all points in the image space. In this case, the AM algorithm converges to the infimum. If the step sizes are not appropriately chosen, then the four-point property is not satisfied for all points in the image space. This necessitates the implementation method to update the image in such a manner that the update points always lies in the subset of the image space where both the properties are satisfied.

In the Chaps. 4–6 our focus was on overcoming the tendency of joint MAP to give trivial solutions by choosing an appropriate PSF regularizer and regularization factor, and the convergence analysis of the resulting optimization problem. In Chap. 7

we provide an alternate way to avoid the trivial solution in MAP by using an image regularizer that has a cost which increases with the amount of blur. We define one such regularizer which is sparsity based and is an improvement over the ℓ_1/ℓ_2-norm of derivative proposed in [76]. We derive a condition to check the suitability of the ratio of ℓ_1-norm of derivative to ℓ_2-norm of higher order derivatives as image priors. We note that the prior proposed in [76] does not exhibit uniform behaviour for all types of images. We classify the images into three types – Type 1, Type 2a, Type 2b, depending on the sparsity of the gradient map and on the strength of the edges. It is noted that only for Type 1 images which have strong edges, the prior proposed in [76] exhibits a monotonic rise in cost function. The regularizer that we have defined exhibits the desired behavior for all types of images including those with sparse and weak edges. We also look into the wavelet domain which is another sparse domain and do deconvolution in this domain using the sparsity enforcing ℓ_1/ℓ_2-norm. Chapter 8 concludes the book and also gives directions for future work.

Chapter 2
Mathematical Background

In this chapter we briefly review concepts from the theory of ill-posed problems, operator theory, linear algebra, estimation theory and sparsity. These concepts are used in the later chapters while doing convergence analysis of alternate minimization for blind image deconvolution and also for arriving at the appropriate regularizers for the image and the point spread function. We have stated theorems where needed without the proof.

We observed in Chap. 1 that under the assumption of a linear shift invariant blur, the image formation model consists of a convolution with a point spread function (PSF) and noise addition. The convolution was assumed to be circular, and the image being two dimensional, the convolution matrix is block circulant with circulant blocks. The convergence analysis in the later chapters proceed by using some properties of the BCCB matrices and these are briefly mentioned in Sect. 2.1. In the previous chapter we pointed out that the inherent difficulty of the blind deconvolution problem stems from the fact that it is an ill-posed problem. Understanding ill-posedness requires concepts from functional analysis. Section 2.2 on operator theory covers the concepts required for the analysis of ill-posed problems. In the next section we look at the different types of ill-posed problems and also methods of handling ill-posedness. We also show that blind deconvolution is a bilinear ill-posed problem. In addition to regularization based methods for blind deconvolution statistical estimation methods are also employed in solving the blind image deconvolution problem. The statistical estimation methods, maximum likelihood and maximum a posteriori probability are described next. Since the derivative domain and transform domain representation of images are sparse in nature, working in these domains gives the advantage of reduced computation owing to the sparse nature. We look into the feasibility of using sparsity based concepts for arriving at an appropriate regularizer. Basic concepts of sparsity are explained in Sect. 2.5. The chapter concludes with an overview of the optimization techniques used in this monograph.

© Springer International Publishing Switzerland 2014
S. Chaudhuri et al., *Blind Image Deconvolution: Methods and Convergence*,
DOI 10.1007/978-3-319-10485-0_2

2.1 Circulant Matrices

For analysis purpose we have taken the convolution in the image formation process to be circular throughout this monograph. It is seen from the image formation equation that under the assumption of circular convolution the PSF matrix obtained is block circulant with circulant blocks (BCCB). We briefly review a few relevant results on circulant matrices [31, 58] in this section.

Definition 2.1. A circulant matrix of order n is a square matrix where each row is obtained by circularly shifting the previous row by one place to the right.

An example is given below:

$$C = \begin{pmatrix} c_1 & c_2 & \cdots & c_n \\ c_n & c_1 & \cdots & c_{n-1} \\ \vdots & & & \\ c_2 & c_3 & \cdots & c_1 \end{pmatrix}.$$

The convolution matrix that arises in the case of 1-dimensional systems is circulant in nature. But for 2-dimensional signals the convolution matrix structure is not circulant but block circulant with circulant blocks which is defined next. A matrix C is block circulant if each row contains blocks which are circularly shifted. Let $C_1, C_2, \cdots, C_m$ be square matrices of order n. A block circulant matrix of order mn is of the form:

$$C = \begin{pmatrix} C_1 & C_2 & \cdots & C_m \\ C_m & C_1 & \cdots & C_{m-1} \\ \vdots & & & \\ C_2 & C_3 & \cdots & C_1 \end{pmatrix}.$$

A matrix in which each element is an order n block can exhibit circulant nature in an alternate manner – each of the blocks is circulant in nature. Such a matrix is called as a matrix with circulant blocks. It is of the form:

$$C = \begin{pmatrix} C_{11} & C_{12} & \cdots & C_{1m} \\ C_{21} & C_{22} & \cdots & C_{2m} \\ \vdots & & & \\ C_{m1} & C_{m2} & \cdots & C_{mm} \end{pmatrix},$$

where each C_{ij} is of order n and is circulant. Having defined a block circulant matrix and a matrix with circulant blocks, we now define a BCCB matrix.

Definition 2.2. A matrix is block circulant with circulant blocks (BCCB) if it is both block circulant and each block is circulant in nature.

Definition 2.3. A matrix C is unitary if

$$CC^{*T} = C^{*T}C = I,$$

i.e., the inverse of a unitary matrix is same as its conjugate transpose.

In the two definitions which follow, an $M \times N$ matrix

$$A = \begin{pmatrix} a(1,1) & \cdots & a(1,N) \\ \vdots & & \vdots \\ a(M,1) & \cdots & a(M,N) \end{pmatrix}$$

is represented as,

$$A = \{a(m,n)\} \qquad 1 \le m \le M, \, 1 \le n \le N.$$

Definition 2.4. The discrete Fourier transform (DFT) matrix of order m is defined as

$$F_m = \left\{ \frac{1}{\sqrt{m}} W_m^{kn} \right\} \qquad 0 \le k, n \le m - 1,$$

where

$$W_m = e^{-j\frac{2\pi}{m}}.$$

Definition 2.5. Given two matrices A of size $M_1 \times M_2$ and B of size $N_1 \times N_2$, their Kronecker product is defined as

$$A \otimes B = \{a(m,n)B\},$$

$$= \begin{pmatrix} a(1,1)B & \cdots & a(1,M_2)B \\ \vdots & & \vdots \\ a(M_1,1)B & \cdots & a(M_1,M_2)B \end{pmatrix}.$$

Having defined the DFT matrix we come to an important theorem regarding BCCB matrices which is used extensively in this monograph.

Theorem 2.1. *All BCCB matrices are diagonalizable by the unitary matrix* $F_m \otimes F_n$.

F_m is the DFT matrix of order m and $\otimes$ is the Kronecker product. If C is a BCCB matrix, then

$$C = (F_m \otimes F_n)^* \Lambda (F_m \otimes F_n), \tag{2.1}$$

where Λ is a diagonal matrix containing the eigenvalues of C which is same as the DFT of the first column of C. From Theorem 2.1 it follows that,

1. The set of BCCB matrices is closed under matrix addition and multiplication. This result is straightforward for matrix addition and for multiplication it follows since the DFT matrix is unitary.
2. Multiplication of BCCB matrices is commutative. This result again follows from the unitary nature of DFT matrix and from the fact that product of diagonal matrices commute.

2.2 Operator Theory

In this section we look at theorems from operator theory needed to understand the nature of ill-posed problems. The theorems are stated without proofs, as they can be found in any standard book on functional analysis [46, 74, 93].

Let $\mathbf{X}$ and $\mathbf{Y}$ be Hilbert spaces and $\mathscr{A}$ a linear operator, $\mathscr{A} : \mathbf{X} \to \mathbf{Y}$. $\mathscr{A}$ is bounded if there is an $\alpha > 0$ such that

$$\| \mathscr{A}(x) \| \leq \alpha \| x \|, \qquad x \in \mathbf{X}. \tag{2.2}$$

The norm in Eq. (2.2) is the norm induced by the inner product in the Hilbert space, i.e., $\| x \| = \langle x, x \rangle^{1/2}$.

Theorem 2.2. *$\mathscr{A}$ is continuous if and only if $\mathscr{A}$ is bounded.*

Definition 2.6. Let $\mathbf{H}$ be a Hilbert space and $\mathscr{A}$ a bounded linear map. The unique bounded linear map $\mathscr{A}^*$ such that

$$\langle \mathscr{A} x, y \rangle = \langle x, \mathscr{A}^* y \rangle$$

for all $x, y \in \mathbf{H}$ is called the adjoint of $\mathscr{A}$.

An operator is self adjoint if $\mathscr{A} = \mathscr{A}^*$.

Definition 2.7. Let $\mathbf{H}$ be a Hilbert space. A linear operator $\mathscr{A}$ is compact if for every bounded sequence (x_n) in $\mathbf{H}$, the image sequence $\mathscr{A}(x_n)$ has a convergent sub-sequence.

$\mathscr{A}$ being compact implies that $\mathscr{A}$ is bounded, the converse is not true in general. A compact operator $\mathscr{A}$ cannot have a bounded (hence continuous) inverse unless its range has finite dimension. This is a result of the following two theorems [46, 93]:

Theorem 2.3. *Let $\mathbf{X}$, $\mathbf{Y}$, and $\mathbf{Z}$ be normed spaces and $\mathscr{K} : \mathbf{X} \to \mathbf{Y}$ and $\mathscr{L} : \mathbf{Y} \to \mathbf{Z}$ be bounded linear operators. Then the product $\mathscr{K}\mathscr{L} : \mathbf{X} \to \mathbf{Z}$ is compact if one of the operators is compact.*

Theorem 2.4. *The identity operator $\mathcal{I} : \mathbf{X} \to \mathbf{X}$ is compact if and only if $\mathbf{X}$ has a finite dimension.*

With $\mathbf{Y} = R(\mathcal{K})$ where $R(\mathcal{K})$ is the range of the operator $\mathcal{K}$ and $\mathbf{Z} = \mathbf{X}$, $\mathcal{L}$ becomes the inverse of $\mathcal{K}$. That is, $\mathcal{L}\mathcal{K} = \mathcal{K}^{-1}\mathcal{K} : \mathbf{X} \to \mathbf{X}$. Now $\mathcal{K}^{-1}\mathcal{K} = \mathcal{I}$. Since $\mathcal{K}$ is compact, for infinite dimensional $\mathbf{X}$, $\mathcal{K}^{-1}$ cannot be bounded. If it were, then it would imply from Theorem 2.3 that $\mathcal{I}$ is compact for infinite dimensional $\mathbf{X}$ which is not true. Hence the inverse operator is bounded (continuous) only when $\mathbf{X}$ is finite dimensional.

The spectral theorem for compact self-adjoint operators is used for the analysis of linear ill-posed problems and we define it next. The spectrum of a bounded linear operator $\mathcal{A}$ in Hilbert space $\mathbf{H}$ is defined as [46, 93]

$$\sigma(\mathcal{A}) = \{\lambda \in \mathbb{C} \,|\, \mathcal{A} - \lambda I \text{ has no bounded inverse}\}, \qquad (2.3)$$

I is the identity operator on $\mathbf{H}$. If $\mathcal{A}$ is self-adjoint then $\sigma(\mathcal{A})$ is a non-empty set of real numbers. For an operator $\mathcal{A}$, if $\mathcal{A}x = \lambda x$, $\lambda \in \mathbb{C}$, then λ is called as an eigenvalue and x an eigenvector associated with the eigenvalue λ. Every eigenvalue is an element of $\sigma(\mathcal{A})$. If $\mathcal{A}$ is self-adjoint then the eigenvectors associated with distinct eigenvalues are orthogonal.

If the operator is compact in addition to being self-adjoint, each non-zero member of the spectrum is an eigenvalue of the operator. The spectrum contains the element zero if the space is infinite dimensional and the sequence of eigenvalues $\lambda_1, \lambda_2, \cdots$ converges to zero for the infinite dimensional case. Also, for each non-zero eigenvalue (λ) of a compact self-adjoint operator, the associated eigenspace (which is the nullspace of the operator $\mathcal{A} - \lambda I$) is finite dimensional. Repeating each eigenvalue in the sequence of eigenvalues $\lambda_1, \lambda_2, \cdots$ according to the dimension of its associated null-space, and listing out the eigenvectors, we get a sequence of orthonormal vectors. Using this the spectral theorem for a compact self-adjoint operator can be stated as [46, 93],

Theorem 2.5. *Let $\mathcal{A} : \mathbf{H} \to \mathbf{H}$ be compact self-adjoint linear operator. Let $\lambda_1, \lambda_2, \cdots$ be the eigenvalues (repeated according to the dimension of the associated eigenspace) and $u_1, u_2, \cdots$ be the associated eigenvectors. Then for any $x \in \mathbf{H}$*

$$\mathcal{A}(x) = \sum_n \lambda_n \langle x, u_n \rangle.$$

The summation is finite or infinite depending on the number of eigenvalues. If A has finitely many eigenvalues then it is said to be of finite rank and corresponds to degenerate kernels for the space of square integrable functions.

2.3   Ill-posed Problems

In Chap. 1 we saw what makes the blind image deconvolution problem ill-posed.
We repeat the definition of ill-posed problem for convenience of the reader. A
problem is well-posed in the Hadamard sense [108] if the problem has a solution
which is unique and if any small perturbation in the data leads to a similar change
in the solution. This last condition indicates the stability of the solution to data
perturbations. A problem which is not well-posed is an ill-posed one. In this section
we see how to handle ill-posedness for the case of linear and non-linear ill-posed
problems.

Let $\mathbf{X}$ and $\mathbf{Y}$ be Hilbert spaces and $\mathscr{K}$ be a transformation from $\mathbf{X}$ to $\mathbf{Y}$.

$$\mathscr{K} : \mathbf{X} \to \mathbf{Y}. \tag{2.4}$$

A problem of the form

$$\mathscr{K}(x) = y, \qquad x \in \mathbf{X}, \, y \in \mathbf{Y}, \tag{2.5}$$

is called an inverse problem when the objective is to determine x, given y or its
perturbed version y^{δ}

$$\| y - y^{\delta} \|^{2} \leq \delta, \tag{2.6}$$

where δ can be thought of as the noise strength and is a small positive quantity.
If the nature of $\mathscr{K}$ is such that some information is lost in the forward process of
formation of y from x, then the inverse problem of determining x from y becomes
an ill-posed problem. Depending on whether $\mathscr{K}$ is a linear or non-linear operator,
the ill-posed problem is classified as linear or non-linear, respectively.

Assuming that the problem of existence and uniqueness of solution is appropriately handled, we will see how to handle the problem of stability of solution using
the concept of regularization. We show in the next section that the ill-posedness of
the blind deconvolution problem arises from the discretized image formation model
and is because of the large condition number of the convolution matrix. With a
large condition number the matrix is close to being singular and the inverse tends
to amplify the noise. Regularizers provide a family of functions that approximate
the inverse of a singular or close to singular operator [46, 50]. In other words, the
ill-posed problem is replaced by a family of well-posed problems whose solutions
are called as the regularized solutions that approximate the original solution.

In the following section we analyze the reason for the ill-posedness of blind
deconvolution. The analysis is based on linear algebra. In the two sections which
follow we show how regularization is used to obtain an approximate solution for the
case of linear and non-linear ill-posed problems in an operator theory framework.

2.3.1 Blind Deconvolution as an Ill-posed Problem

In this section we see what makes the blind deconvolution problem inherently difficult to solve. Deconvolution when the PSF is known itself is a difficult problem to solve owing to the fact that it is an inverse problem. In most of the inverse problems the difficulty in solution arises due to the fact that some information is lost in the forward process, which makes inversion a difficult task. In order to get to the root of the problem, let us look at the image formation process which is an analog process. Under the assumptions of linearity and shift invariance, the image formation equation in the continuous domain is given by Eq. (1.6), which is repeated here for convenience.

$$y(u,v) = \int_{-\infty}^{\infty} \int_{-\infty}^{\infty} k(u - u_1, v - v_1)x(u_1, v_1)du_1dv_1.$$

In image restoration $x(u,v)$ needs to be estimated from a noisy version of $y(u,v)$. This amounts to solving the integral equation in Eq. (1.6), which is a Fredholm integral equation of the first kind, the general form of which in one dimension is given by

$$\int_{a}^{b} \mathscr{K}(u,v)x(v)dv = y(u), \qquad (2.7)$$

where $x(.)$ is the unknown function, which is to be solved for, $\mathscr{K}(u,v)$ is the kernel and $y(u)$ is a known function.

It can be shown that an integral operator of the type in Eq. (2.7) is a compact operator by proving that it is a Hilbert-Schmidt operator [93].

Definition 2.8. Let H be a Hilbert space and $\mathscr{A} : X \rightarrow Y$ be a linear operator. $\mathscr{A}$ is called a Hilbert-Schmidt operator if there is an orthonormal basis (v_n) of H such that

$$\sum_{n} \| \mathscr{A}(v_n) \|^2 < \infty.$$

If a bounded linear operator is Hilbert-Schmidt then it is compact [93]. It is a well known fact that [74, 93, 110],

Theorem 2.6. *Let X and Y be normed spaces and let $\mathscr{A} : X \rightarrow Y$ be a compact linear operator. $\mathscr{A}$ cannot have a bounded inverse if X is infinite dimensional.*

It may be noted that Theorem 2.6 follows from Theorems 2.3 and 2.4. From Theorem 2.6 it is inferred that the inverse operator of Eq. (2.7) is not continuous. Since Eq. (1.6) is of the form Eq. (2.7), the inverse operator in this case is also discontinuous. This leads to instability of the solution making the problem an ill-posed one. An alternate way to see the instability in solution of the integral equation in Eq. (2.7) is by making use of the Reimann-Lebesgue lemma [124]. From Riemann-Lebesgue lemma, for any integrable kernel, the integral

$$\int_a^b \mathcal{K}(u,v)\sin(mv)dv \to 0 \text{ as } m \to \infty,$$

and the integral tends to zero faster for smooth kernels than for sharp ones. If $x(v)$ is a solution to Eq. (1.6), then

$$\int_a^b \mathcal{K}(u,v)(x(v)+\sin(mv))dv = y(u) + \int_a^b \mathcal{K}(u,v)\sin(mv)dv. \qquad (2.8)$$

As $m \to \infty$, the perturbation in y is infinitesimal, whereas there is a finite change for x, indicating instability – the solution is sensitive to high frequency perturbations. Since in image deconvolution we work in the finite dimensional discrete domain, the problem of discontinuous inverse does not exist since in this case a compact linear operator always has a continuous inverse. Yet, since the finite dimensional case is obtained by discretizing an ill-posed problem, the effect of ill-posedness of the continuous operator on the behavior of the discrete problem needs to be explored.

For the image formation model in Eq. (1.5), discretization gives an expression of the form Eq. (1.9) with the noise term being zero. Here K is the finite dimensional approximation of the kernel $\mathcal{K}$. It has been noted [9, 73, 124] that, instability in solution of the continuous domain formulation Eq. (1.5) gets reflected in the condition number of its finite dimensional approximation. It is observed that the condition number depends on the quality of the discretization. Finer the quality of discretization, higher is the condition number aggravating the stability issues while doing the image restoration. An explanation for this behavior is given in [128]. Using the fact that the convolution matrix K is the discretized version of the kernel $\mathcal{K}$ and using Taylor's theorem, it was shown in [128] that, any row of K is approximately a linear combination of the adjacent rows leading to a nearly singular matrix. It is also shown that, finer the discretization, better the approximation, which makes K more ill-conditioned.

It may be noted that though the continuous domain formulation is ill-posed due to stability issues, the equivalent discrete domain problem is not ill-posed strictly in the sense of the third condition for ill-posedness. If the discretization is coarse, then the condition number is small and a solution can still be obtained, but as the fineness of the discretization increases, the behavior of the discrete problem approximates the unstable behavior of the continuous equivalent, in the sense that it becomes sensitive to high frequency perturbations. The effect of the condition number on the usability of the estimated solution is analyzed next.

The convolution matrix K being BCCB, is diagonalized by the DFT matrix. i.e.

$$K = F^*\mathcal{D}F, \qquad (2.9)$$

where F is the 2D DFT matrix, F^* its conjugate transpose which is same as the inverse in this case (i.e. $F^{-1} = F^*$) since the DFT matrix is unitary. $\mathscr{D}$ is a diagonal matrix with diagonal entries corresponding to the DFT of the first column of K, which is also same as the eigenvalues of K. Let the columns of F^* be represented by v_k, $k = 0, 1, \cdots, N$, then Eq. (2.9) becomes

$$K = \sum_{k=1}^{N} \lambda_k v_k v_k^*, \tag{2.10}$$

where λ_k are the eigenvalues of K. v_k^* is the conjugate transpose of v_k. For noisy observations, from Eq. (1.9)

$$\underline{y} = K\underline{x} + \underline{n},$$

$$= \sum_{k=1}^{N} \lambda_k v_k v_k^* \underline{x} + \underline{n}. \tag{2.11}$$

where Eq. (2.10) is used for K. Since the PSF is usually a low pass filter, the high frequency spectral content is low which corresponds to small eigenvalues. For very low eigenvalues it may be seen from Eq. (2.11) that the noise dominates over the signal and for an image this signal component corresponds to the edges. Provided K^{-1} exists, a possible estimate for the image ($\hat{x}$) is

$$\hat{\underline{x}} = K^{-1}\underline{y},$$

$$= \underline{x} + K^{-1}\underline{n}. \tag{2.12}$$

From Eqs. (2.9) and (2.10)

$$K^{-1} = F^* \mathscr{D}^{-1} F,$$

$$= \sum_{k=1}^{N} \frac{v_k v_k^*}{\lambda_k}. \tag{2.13}$$

Using Eq. (2.13), in Eq. (2.12) we get

$$\hat{\underline{x}} = \underline{x} + \sum_{k=1}^{N} \frac{v_k (v_k^* \underline{n})}{\lambda_k}. \tag{2.14}$$

From Eq. (2.14) we see that even though the original image is extracted by the inverse, there is a noise component added to this. For eigenvalues close to zero, the noise amplification is very high making the estimate unusable. The ill-posed nature is evident here, since a small change in the data leads to a large variation in the estimated image. Let the error in the data be upper bounded by δ, i.e.

$$\| \underline{y}^\delta - \underline{y} \| \le \delta, \tag{2.15}$$

where $\underline{y}^\delta$ is the perturbed observation. From Eq. (2.14), the corresponding error in the estimate is

$$\| \hat{\underline{x}} - \underline{x} \|^2 = \sum_{k=1}^{N} \frac{|v_k^*(\underline{y}^\delta - \underline{y})|^2}{\lambda_k^2}. \tag{2.16}$$

Let the eigenvalues be arranged such that $0 \le |\lambda_1| \le \cdots \le |\lambda_N|$. It can be seen that

$$\text{If } \underline{y}^\delta - \underline{y} = \delta v_1, \text{ then } \| \hat{\underline{x}} - \underline{x} \| = \delta/|\lambda_1|$$

$$\text{If } \underline{y}^\delta - \underline{y} = \delta v_N, \text{ then } \| \hat{\underline{x}} - \underline{x} \| = \delta/|\lambda_N|.$$

This implies that for low frequencies (large eigenvalues), the effect of noise on the reconstructed signal is very less, whereas for high frequencies the signal is severely corrupted by noise. This is characteristic of ill-posed problems – the same amount of noise in different frequency bands is treated differently. It may also be noted that λ_1, the smallest eigenvalue determines the condition number of a matrix which is defined as

$$\kappa = \frac{|\lambda_N|}{|\lambda_1|}. \tag{2.17}$$

As λ_1 tends to zero, the matrix becomes more ill-conditioned leading to instability of the solution. These facts can also be inferred from a frequency domain analysis. A complete frequency domain analysis of the problem can be found in [9]. Since the results w.r.t. stability are same, we provide only a summary of the existence and uniqueness conditions here. The frequency domain equivalent of Eq. (1.6) in the presence of additive noise is

$$\mathscr{Y}(\Omega) = \mathscr{K}(\Omega)\mathscr{X}(\Omega) + \mathscr{N}(\Omega), \tag{2.18}$$

where $\mathscr{Y}$, $\mathscr{X}$, and $\mathscr{K}$ denote the Fourier transform for the continuous case, and $\Omega = [\Omega_x, \Omega_y]$. Since the system is linear, the condition for uniqueness of solution is

$$\mathscr{K}(\Omega)\mathscr{X}(\Omega) = 0, \tag{2.19}$$

only when $\mathscr{X}(\Omega) = 0$. If the system is band limited, this is not true and a unique solution does not exist. Assuming that $\mathscr{K}(\Omega)$ has the entire frequency axis as its support, we can look at the condition for existence of solution. As in Eq. (2.14) in spatial domain, a estimate in the frequency domain is obtained as

$$\hat{\mathscr{X}}(\Omega) = \mathscr{X}(\Omega) + \frac{\mathscr{N}(\Omega)}{\mathscr{K}(\Omega)}. \tag{2.20}$$

Now, the existence of solution depends on the existence of inverse Fourier transform of Eq. (2.20). For the case of motion blur, the frequency response of the imaging system is zero for some frequencies, whereas the noise spectrum is assumed to be uniform through the spectrum (under the assumption of white noise). This leads to singularities at the zeros of the PSF and the inverse of $\hat{\mathscr{X}}(\Omega)$ may not exist. Even if $\mathscr{K}(\Omega)$ is not zero, it tends to zero as $\Omega \to \infty$, but the ratio $\dfrac{\mathscr{N}(\Omega)}{\mathscr{K}(\Omega)}$ may not tend to zero. Hence the existence of the inverse Fourier transform depends on the behavior of this ratio, and the condition for existence of inverse can be expressed as,

$$\int \left| \frac{\mathscr{N}(\Omega)}{\mathscr{K}(\Omega)} \right| d\Omega < \infty. \tag{2.21}$$

In general Eq. (2.21) does not hold for noisy images, and hence the solution exists only for noise free images. The uniqueness and existence in the discrete case can also be determined on similar lines. The above analysis is for the case where the PSF is known. In the case of blind deconvolution the problem is more complicated due to the fact that the PSF is not known. This makes the problem highly nonlinear, and there exists an infinite number of solutions of the form $(a\hat{\underline{x}}, \ \hat{k}/a)$, where $a \in \mathbf{R}^+$. Though the solution space can be reduced by using the constraints on PSF and using known properties of the image, the problem of stability still persists. Next we see the two different classes of ill-posed problems namely the linear and non-linear ill-posed problems.

2.3.2 Linear Ill-posed Problems

For the linear case, $\mathscr{K}$ in Eq. (2.4) is a bounded linear operator from $\mathbf{X}$ to $\mathbf{Y}$. In the continuous case, the operator is denoted by $\mathscr{K}$ and its discretized version which is a matrix is written as K. Due to the ill-conditioned nature of the problem we look for a vector which is 'near' to the solution instead of an exact solution. The best approximate solution $\hat{x}$ for $Kx = y$ is taken as the minimum-norm least squares solution [33] i.e.,

$$\| \hat{x} \| = \inf \{\| Kx - y \| \mid x \in X\}. \tag{2.22}$$

It is proved in [33] that there exists a unique best approximate solution given by

$$\hat{x} = K^\dagger y, \tag{2.23}$$

where $K^\dagger$ is the Moore-Penrose generalized inverse. Also $x \in X$ is a least squares solution if and only if

$$K^* K x = K^* y, \tag{2.24}$$

i.e., $K^\dagger = (K^*K)^{-1}K^*$, where K^* is the adjoint of K and for real valued K the adjoint K^* reduces to the transpose K^T.

As mentioned in the Sect. 2.2, if K is compact and the range of K is infinite dimensional then $K^\dagger$ is unbounded and hence discontinuous leading to illposedness. From the spectral theorem, for a compact self-adjoint linear operator there exists a singular system defined as $(\sigma_n; v_n, u_n)$ [46, 93], where σ_n^2 are the non-zero eigenvalues of the operator K^*K, written in decreasing order, $\{v_n\}$ are the corresponding orthonormal system of eigenvectors of K^*K, and $\{u_n\}$ are the orthonormal system of eigenvectors for KK^*. The condition for existence of a best approximate solution is given by Picard criterion [33], which says that such a solution exists if the Fourier coefficients ($\langle y, u_n \rangle$) decay fast enough relative to the singular values σ_n. For compact linear operators whose range is infinite dimensional the eigenvalues tend to zero as $n \to \infty$ which leads to noise amplification as we saw in the previous section. Faster the eigenvalues decay, stronger is the error amplification. Using this fact, ill-posedness can be quantified as mild or severe depending on polynomial or exponential decay, respectively, of the eigenvalues.

Unboundedness of $K^\dagger$ leads to ill-posedness since the inverse is not continuous in this case and this creates problem when the data is noisy. Let y^δ be the noisy observation with

$$\| y^\delta - y \| \le \delta, \tag{2.25}$$

where δ is the noise variation. Owing to the discontinuity of $K^\dagger$, $K^\dagger y^\delta$ is not a good approximation to $K^\dagger y$. This necessitates finding an approximation to the best approximate solution $\hat{x} = K^\dagger y$ when the data is noisy. To handle instability it is required to find a continuous function that approximates $K^\dagger$ and gives a solution x_α^δ which converges to $\hat{x}$ when $\delta \to 0$ with a proper choice of α. This is achieved by replacing $K^\dagger$ by a family $\{R_\alpha\}$ of continuous operators referred to as regularization operators, which depend on the parameter α (the regularization factor). With this

$$x_\alpha^\delta = R_\alpha y^\delta, \qquad x_\alpha^\delta \to x^\dagger, \text{ when } \delta \to 0. \tag{2.26}$$

A family of operators $\{R_\alpha\}$ exists since there exists a collection of equations, $Kx = y^\delta$, corresponding to different values of δ. Fixing a value for α depends on a specific equation due to its dependency on δ. In the discrete case, the requirement of a continuous inverse is equivalent to the condition number of the matrix equivalent of $K^\dagger$ being small. We describe two types of regularizers below – the Tikhonov regularizer and the bounded variation regularization. Since we focus only on estimation of the discrete image, these regularizers are explained in the discrete domain.

2.3.2.1 Tikhonov Regularization

We observed in Sect. 2.3.1 that in the presence of noise, oscillations appear in the solution of Fredholm integral equation. Tikhonov [50, 122, 157] proposed usage of an additional term to smooth out such oscillations. A spectral theory based development [33, 46] of the Tikhonov regularizer is given below.

Given the singular system $(\sigma_n; v_n, u_n)$ for the compact linear operator K, $K^\dagger y$ can be written as

$$K^\dagger y = \sum_{n=1}^{\infty} \frac{\langle K^* y, v_n \rangle}{\sigma_n^2} v_n. \qquad (2.27)$$

To regularize $K^\dagger$ the amplification factors $\dfrac{1}{\sigma_n^2}$ is replaced by a modified version $U(\alpha, \sigma_n^2)$ [33]. This gives

$$x_\alpha^\delta = \sum_{n=1}^{\infty} U(\alpha, \sigma_n^2) \langle K^* y^\delta, v_n \rangle v_n. \qquad (2.28)$$

The choice of $U(\alpha, \sigma_n^2)$

$$U(\alpha, \sigma_n^2) = \frac{1}{\alpha + \sigma_n^2}, \qquad (2.29)$$

leads to the Tikhonov regularizer. This, along with $K v_n = \sigma_n u_n$ gives

$$x_\alpha^\delta = \sum_{n=1}^{\infty} \frac{\sigma_n}{\alpha + \sigma_n^2} \langle y^\delta, u_n \rangle v_n. \qquad (2.30)$$

In terms of the operator K, x_α^δ given in Eq. (2.30) can be written as

$$x_\alpha^\delta = (\alpha I + K^* K)^{-1} K^* y^\delta. \qquad (2.31)$$

This is the solution obtained by minimizing the cost

$$C_\alpha^\delta(x) = \| Kx - y^\delta \|^2 + \alpha \| x \|^2 . \qquad (2.32)$$

The presence of $\alpha \| x \|^2$ stabilizes the solution. It penalizes solutions which are of large amplitude, possibly due to error amplification. Penalty terms of a more general form such as $\| \mathscr{L}x \|^2$ could be used where $\mathscr{L}$ is usually a derivative based operator that enforces smoothness. With $\| \mathscr{L}x \|^2$ as the regularizer the estimate becomes

$$C_\alpha^\delta(x) = (\alpha \mathscr{L}^* \mathscr{L} + K^* K)^{-1} K^* y^\delta. \qquad (2.33)$$

To see the effect of the regularization parameter α, we split Eq. (2.30) by using the image formation model $y^\delta = Kx + n$, where n is AWGN. This model is same as in Eq. (1.9), only the notation is changed.

$$x_\alpha^\delta = \sum_{n=1}^\infty \frac{\sigma_n^2}{\alpha + \sigma_n^2} \langle x, v_n \rangle v_n + \sum_{n=1}^\infty \frac{\sigma_n}{\alpha + \sigma_n^2} \langle n, u_n \rangle v_n. \tag{2.34}$$

It is seen from Eq. (2.34) that for small values of α the first term is approximately x and the second term leads to noise amplification since the denominator is dominated by σ_n^2. This amounts to having no regularization. If α is large, the effect of noise amplification is less, but the solution deviates from x. Hence the regularization factor should be selected such that there is a balance between stability and accuracy.

The problem with using a regularization of the form Eq. (2.33) in image restoration is that the solution is excessively smooth since the edges in the image are penalized by the regularizer. Since $\mathscr{L}$ is a derivative operator its eigenvalues increases as $n \to \infty$. With $\mathscr{L}x$ as the regularizer, the solution becomes

$$x_\alpha^\delta = \sum_{n=1}^\infty \frac{\sigma_n^2}{\alpha \gamma_n^2 + \sigma_n^2} \langle x, v_n \rangle v_n + \sum_{n=1}^\infty \frac{\sigma_n}{\alpha \gamma_n^2 + \sigma_n^2} \langle n, u_n \rangle v_n, \tag{2.35}$$

where γ_n is the eigenvalue of the differential operator γ_n. Differential operators are unbounded and in the case where they have a compact self-adjoint inverse, the eigenvalues of the operator is the reciprocal of the eigenvalues of the inverse operator. As was mentioned earlier the eigenvalues of a compact self-adjoint operator tends to zero as $n \to \infty$, when there are infinite of them. Assuming that $\mathscr{L}$ has a compact self-adjoint inverse, γ_n increases as $n \to \infty$. Since γ_n increases and σ_n decreases with n it can be seen from Eq. (2.35) that the amount of regularization depends on n which corresponds to frequency in the signal processing perspective. For small frequencies since γ_n is close to zero, the signal is reproduced faithfully. The noise component gets attenuated more as frequency increases since the denominator of the second term increases with n. With increase in n it is also seen that the first term gets attenuated more leading to loss of high frequency information in x which corresponds to edge information in image, leading to smoothing of the image. It may be noted that the regularizer corresponds to a high-pass filter and the blur term to a low-pass filter in a signal processing perspective. This aspect is further discussed in Chap. 3. In order to avoid smoothing of the solution we need to use a regularizer which permits edges in the solution. Total variation which is one such regularizer, is described next.

2.3.2.2 Bounded Variation Regularization

A function of bounded variation is one with a finite total variation. For a continuous function $f(u)$, the total variation is defined as [137]

$$TV(f(u)) = \int_a^b |f'(u)| du. \tag{2.36}$$

This measures the amount by which the function changes along the ordinate as the value of the independent variable changes from a to b. For a differentiable function of two variables $f(u, v)$, total variation becomes

$$TV(f(u, v)) = \int_a^b \int_c^d |\text{ grad } f(u, v)| du dv, \tag{2.37}$$

where grad f is the gradient of the function f and $|\text{ grad } f(u, v)|$, the magnitude of the gradient is $\sqrt{f_u^2 + f_v^2}$, with f_u, f_v being the first order horizontal and vertical derivatives, respectively. The isotropic discrete total variation function for an image [8, 15] is defined as

$$TV(x) = \sum_j \sqrt{\Delta^h(x_j)^2 + \Delta^v(x_j)^2}, \tag{2.38}$$

where Δ^h and Δ^v are the first-order horizontal and vertical differences, respectively, and j is the pixel index. It can be seen as the sum of magnitude of the gradient at each pixel location. $TV(x)$ can be seen as a mixed norm, since at each pixel location an ℓ_2 norm (of the gradient at that point) is taken, followed by summation of the positive square root values which amounts to an ℓ_1 norm. Replacing the stabilizing term of Eq. (2.32) with Eq. (2.38), leads to total variation regularization which permits solutions that can capture sharp features in the original image.

Unlike the Tikhonov regularizer, here the regularization term is non-quadratic. In order to make analysis feasible we approximate the TV function by its quadratic upper bound [40]. Since Eq. (2.38) is the sum of square root of a quadratic function, an obvious way to upper bound $TV(x)$ is by upper bounding the square root function [40]. This is illustrated in Fig. 2.1. From Fig. 2.1

$$\sqrt{x_2} \leq \sqrt{x_1} + \frac{1}{2\sqrt{x_1}}(x_2 - x_1), \qquad x_2 \geq x_1, x_1 > 0. \tag{2.39}$$

where $\frac{1}{2\sqrt{x_1}}$ is the slope of $\sqrt{x}$ at the point x_1. Using Eq. (2.39) in Eq. (2.38)

$$TV(x) \leq Q_{TV}(x, x^{(i)}),$$

$$\triangleq \sum_j \frac{1}{2} \frac{(\Delta^h x_j)^2 + (\Delta^v x_j)^2}{\sqrt{(\Delta^h x_j^{(i)})^2 + (\Delta^v x_j^{(i)})^2}} + B(x^{(i)}), \tag{2.40}$$

where i is the iteration number, $x^{(i)}$ is the point w.r.t. which the value of TV function at x is approximated. $B(x^{(i)})$ is a constant term, independent of the

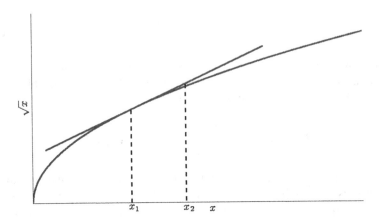

Fig. 2.1 Illustration of upper bounding the square root function

variable x. The right hand side of the inequality Eq. (2.40) is a quadratic function of the first order difference of image pixel values. The first term of Eq. (2.40) can be written as a matrix vector product by defining a difference operation matrix $D = [(D^h)^T \ (D^v)^T]^T$. D^h and D^v denote matrices such that $D^h\underline{x}$ and $D^v\underline{x}$ are the vectors of all horizontal and vertical first-order differences of the vector x, respectively ($\underline{x}$ is the vector obtained by row ordering the matrix x). Periodic boundary condition is used for x while computing the first-order difference. Using the operator D, Eq. (2.40) becomes

$$Q_{TV}(\underline{x}, \underline{x}^{(i)}) = \underline{x}^T D^T \Lambda^{(i)} D \underline{x} + B(\underline{x}^{(i)}), \tag{2.41}$$

where $\Lambda^{(i)}$ takes care of the denominator of the first term in RHS of Eq. (2.40) and is defined as

$$\Lambda^{(i)} = \text{diag}(W^{(i)}, W^{(i)}), \tag{2.42}$$

where $\text{diag}(L)$ refers to a diagonal matrix with elements of vector L as the diagonal and $W^{(i)}$ is a vector whose jth element is

$$w_j^{(i)} = \left(2\sqrt{(\Delta^h x_j^{(i)})^2 + (\Delta^v x_j^{(i)})^2}\right)^{-1}. \tag{2.43}$$

In all our formulations with TV as regularizer we use the upper bounded function in Eq. (2.41) as the regularizer.

2.3.3 Non-linear Ill-posed Problems

An ill-posed problem is classified as non-linear, when $\mathcal{H}$ in Eq. (2.4) is a non-linear operator. Since spectral analysis is infeasible in this case, finding a regularizer and regularization factor is not as straight forward as in the linear ill-posed case. Tikhonov regularization and iterative methods [33,50] are two widely used methods for solving non-linear ill-posed problems. With $\mathscr{K}$ non-linear, Eq. (2.32) can be rewritten as

$$C_\alpha^\delta(x) = \parallel \mathscr{K}(x) - y^\delta \parallel^2 + \alpha \parallel x \parallel^2 . \tag{2.44}$$

and it can be seen that uniqueness of the solution is not guaranteed in this case. With $\mathscr{K}$ non-linear, Eq. (2.44) is not convex and may have local minima in which a descent method could get stuck. Also determination of a proper regularization factor is difficult in this case. Owing to these difficulties iterative regularization methods are preferred for the non-linear case.

In iterative methods x is estimated starting from an initial guess x_0 as

$$x_{i+1} = x_i + \mathscr{K}'(x_i)^{-1}(y - \mathscr{K}(x_i)), \tag{2.45}$$

where $\mathscr{K}'$ is the Frechet derivative of $\mathscr{K}$ [50]. This is similar to the Newton's method, with the difference that the derivative of the function is replaced by the Frechet derivative which is defined below [119, 126].

Definition 2.9. Let $\mathbf{X}$ and $\mathbf{Y}$ be two normed linear spaces. A mapping $\mathscr{F} : \mathbf{X} \to \mathbf{Y}$ is said to be Frechet differentiable at $x \in \mathbf{X}$ if there exists a bounded linear mapping $\mathscr{A} : \mathbf{X} \to \mathbf{Y}$ such that

$$\lim_{h \to 0} \frac{\parallel \mathscr{F}(x+h) - \mathscr{F}(x) - \mathscr{A}h \parallel_\mathbf{Y}}{\parallel h \parallel_\mathbf{X}} = 0.$$

The linear operator $\mathscr{A}$ is called the Frechet derivative of $\mathscr{F}$ at x.

Frechet derivative defines the derivative of an operator and is given by the bounded linear operator $\mathscr{A}$. This concept is similar to that of the derivative of functions, wherein the derivative represents the slope of the tangent at the point and is the best linear approximation in the neighborhood of the point at which the derivative is taken. For ill-posed problems the inverse of $\mathscr{K}'$ is usually unbounded because of which one has to solve a linear ill-posed problem in each step of the iteration by using a regularizer. Hence an iterative method of the form Eq. (2.45) is inappropriate. An alternative is to use Landweber iteration [50] which is a steepest descent method leading to an update equation of the form

$$x_{i+1} = x_i + \mathscr{K}'(x_i)^*(y - \mathscr{K}(x_i)), \tag{2.46}$$

where $\mathscr{K}'(x_i)^*$ is the adjoint of the Frechet derivative. Here the negative gradient of the functional $\frac{1}{2} \parallel y - \mathscr{K} \parallel^2$ (second term in Eq. (2.46)) decides the update direction for the current iteration. In Eqs. (2.45) and (2.46), i indicates the iteration.

Due to the ill-posed nature of the problem, the iterations must be stopped at an appropriate time or stopping index i_*. An appropriate value of i_* can be chosen as a function of the perturbation in the data δ. The index i_* can be determined using

$$\| y^\delta - F(x_{i_*}^\delta) \| \le \tau\delta < \| y^\delta - F(x_i^\delta) \|, \quad 0 \le i < i_*, \tag{2.47}$$

for some sufficiently large $\tau > 0$. This condition ensures that the error is less than $\tau\delta$ when the number of iterations becomes i_*.

2.3.4 Bilinear Ill-posed Problems

In this section we show that the blind deconvolution problem is a bilinear ill-posed problem. Equation (2.5) is similar to the image formation equation (1.9) with $\mathscr{K}$ being the convolution matrix K. In the case of blind deconvolution, the operator K which is the convolution matrix corresponding to the point spread function is not known which may make the blind deconvolution problem look like a non-linear ill-posed problem, since both K and $\underline{x}$ are unknowns. Considering the image and the PSF as unknowns, the image formation process can be seen as mapping of the form

$$\mathscr{G} : \mathbf{X} \times \mathbf{K} \to \mathbf{Y}, \tag{2.48}$$

where $\mathbf{K}$ is the Hilbert space consisting of the PSFs. The image formed $\underline{y}$ can be written in terms of the mapping as

$$\underline{y} = \mathscr{G}(\underline{x}, K),$$
$$= K\underline{x}, \tag{2.49}$$

in the no-noise case. Keeping one of the variables, say K, constant it is seen from Eq. (2.50) that the operator $\mathscr{G}$ is linear in the other variable x

$$\mathscr{G}(\alpha_1\underline{x}_1 + \alpha_2\underline{x}_2, K) = K(\alpha_1\underline{x}_1 + \alpha_2\underline{x}_2)$$
$$= \alpha_1 K\underline{x}_1 + \alpha_2 K\underline{x}_1$$
$$= \alpha_1 \mathscr{G}(\underline{x}_1, K) + \alpha_2 \mathscr{G}(\underline{x}_2, K), \tag{2.50}$$

where α_1 and α_2 are constants. Similarly by fixing $\underline{x}$ and using the constants β_1 and β_2, it is seen that the operator is linear in the PSF variable K

$$\mathscr{G}(\underline{x}, \beta_1 K_1 + \beta_2 K_2) = \beta_1 \mathscr{G}(\underline{x}, K_1) + \beta_2 \mathscr{G}(\underline{x}, K_2). \tag{2.51}$$

From Eqs. (2.50) and (2.51) we conclude that the blind deconvolution process is a bilinear ill-posed problem. Since keeping one of the variables fixed reduces the problem to a linear ill-posed one, solving the blind deconvolution problem amounts to solving two linear ill-posed problems that are coupled. Solution for problems of this type is pursued in Sect. 2.6.

2.4   Estimation of Random Field

In the above sections we considered image to be a deterministic signal and used regularization methods to handle the ill-posedness of blind deconvolution. An alternative to this is to use statistical estimation methods. Here the image is considered to be a stochastic signal, with each pixel represented as a random variable. Such a statistical representation of an image is called a random field [58]. There exists a rich structure in natural images and, in general, the neighboring pixels have similar values except at the edges. This structure is often captured by a Markov random field (MRF) [20, 58]. Given a pixel at location (i, j), let its limited neighborhood be represented by $N(i, j)$. Let X_{ij} be the random variable representing the pixel value at (i, j). A given random field shows the Markov property if

$$P(X_{ij} = x_{ij} | X_{kl} = x_{kl}, \forall (k, l) \neq (i, j)) = P(X_{ij} = x_{ij} | X_{kl} = x_{kl}, \forall (k, l) \in N(i, j)),$$

(2.52)

i.e., the value of a pixel depends only upon the value of its neighboring pixels. With Markov property one can only get the conditional distribution of the random field, whereas one needs the joint distribution of the pixels in the random field to serve as the prior. The joint distribution can be obtained using the Hammersley-Clifford theorem [10] which establishes a one to one correspondence between an MRF and a Gibbs distribution. Before proceeding to the theorem we would define the Gibbs distribution for which one needs to define a clique. Given a neighborhood $N(i, j)$ for a pixel, a clique is either a single pixel location or a pair such that if (i, j) and (k, l) belong to a clique (c), then $(k, l) \in N(i, j)$. We do not discuss higher order cliques here. Let the collection of all cliques c be denoted by C.

Definition 2.10. A random field with a neighborhood structure $N(i, j)$ defined on it is a Gibbs random field *iff* its joint distribution is of the form

$$P(X = x) = \frac{1}{Z} \exp(-U(x)),$$

(2.53)

where $U(x)$ is the energy function associated with all the cliques and is defined as

$$U(x) = \sum_{c \in C} V_c(x),$$

where $V_c(x)$ is the clique potential associated with the clique c and Z is the partition function given by

$$Z = \sum_x U(x).$$

Here the summation is carried out over the entire configuration space of the variable x.

We now state the Hammersley-Clifford theorem without proof, the proof can be found in [10].

Theorem 2.7. *Let $N(i, j)$ be a neighborhood system defined on a random field. The random field is a Markov random field w.r.t. $N(i, j)$ iff its joint distribution is a Gibbs distribution with cliques associated with $N(i, j)$.*

Using this theorem one can obtain a joint distribution for a natural image which can be used as a prior for the image.

In the image formation model in Eq. (1.9) we assumed the noise to be white Gaussian leading to a random vector $\underline{n}$ consisting of Gaussian distributed random variables which are independent with mean zero and variance same as the noise variance (σ_n^2). Hence given the original image $\underline{x}$ and the PSF $\underline{k}$, $\underline{y}$ is a random vector which also consists of independent Gaussian random variables with means $K\underline{x}$ and variance σ_n^2. With this, the likelihood of $\underline{y}$ can be written as

$$f(\underline{y}|\underline{x}, \underline{k}) \propto \exp\left(-\frac{\| \underline{y} - K\underline{x} \|^2}{2\sigma_n^2} \right). \tag{2.54}$$

The a posteriori probability for the image and the PSF is

$$f(\underline{x}, \underline{k}|\underline{y}) = \frac{f(\underline{y}|\underline{x}, \underline{k}) f(\underline{x}) f(\underline{k})}{f(\underline{y})}, \tag{2.55}$$

where the density corresponding to a variable is to be understood from the argument (i.e. $f_x(x)$ is written as $f(x)$). We have assumed that the image and the PSF are independent random vectors. Here $f(x)$ and $f(k)$ are the priors of the image and the PSF, respectively. Since the image and the PSF both can be seen as random fields, following the discussion at the beginning of this section, both can be seen as Markov random fields and a prior can be assigned using the Hammersley-Clifford theorem. Usually this gives rise to the quadratic prior which enforces smoothness in the solution.

In maximum likelihood (ML) estimation, $\underline{x}$, and $\underline{k}$ which maximize the likelihood need to be estimated. Instead of maximizing the likelihood as given in Eq. (2.54), its negative logarithm is minimized.

$$(\hat{\underline{x}}, \hat{\underline{k}}) = \arg\min_{\underline{x}, \underline{k}} \lambda \| \underline{y} - K\underline{x} \|^2, \tag{2.56}$$

where λ replaces the unknown noise variance. Similarly in maximum a posteriori probability (MAP) estimate, the negative logarithm of a posteriori probability is minimized to estimate the image and the PSF which maximize the a posteriori probability. The MAP estimate is given by

$$(\hat{\underline{x}}, \hat{\underline{k}}) = \underset{\underline{x},\underline{k}}{\arg\min} \parallel \underline{y} - K\underline{x} \parallel^2 -\lambda_x \log f(\underline{x}) - \lambda_k \log f(\underline{k}), \tag{2.57}$$

where λ_x and λ_k replace the noise variance. Since both the image and the PSF are unknowns, ML cannot be used directly for estimating the unknowns. But ML is used in blind deconvolution for estimating model parameters in the case where the image and PSF are modeled using ARMA (auto regressive moving average) model; this is given in more detail in Chap. 3. Since in MAP estimation additional information in the form of prior is present it is more suited for blind deconvolution; this is discussed further in Chap. 4. It may also be noted that MAP estimation is like the regularized estimation with the priors acting as regularizers. One way of modeling the blur prior is using parametric models like the linear motion blur, out-of-focus blur and the atmospheric turbulence blur, the details of which were given in Chap. 1. Another way of assigning prior to image and blur is by making use of the structure present in image/blur to restrict the solution space to that of the most probable solutions. This amounts to the regularization method used to convert an ill-posed problem to a well-posed one. Since natural images are smooth/piecewise smooth/textured, this information can be captured in stochastic models and be used as the prior [123]. The priors are specified by taking into consideration the probabilistic relations between neighboring pixels in the image or the PSF as discussed.

2.5 Sparsity Inducing Norms

Sparsity is a suitable concept which can be used for finding appropriate regularizers for an image since images are often sparse in the derivative domain and in the transform domain. In this section we introduce the concept of sparsity and also look at sparsity inducing norms that can be used as potential regularizers.

A signal or vector can be termed as sparse if most of the energy of the signal is concentrated on a few samples and the remaining samples are zero. It is a well known fact that the ℓ_1 norm induces sparsity i.e., using ℓ_1 norm as a regularizer while solving problems of the form $Ax = b$ leads to solutions which have more number of zeros. Reviewing the literature [12,22,32,156] it is seen that this tendency of ℓ_1 norm is accepted as a heuristic and there exists only graphical ways of explaining this behavior. For purpose of completeness we describe the graphical reasoning for sparsity inducing property of ℓ_1 norm mentioned in [117, 156].

With a quadratic data term and a ℓ_1 norm regularizer the cost function is of the form

$$C(x) = \parallel Ax - b \parallel_2^2 + \parallel x \parallel_1 . \tag{2.58}$$

The solution to this lies at the common tangent of the level surface of the ℓ_1 and the ℓ_2 norms [117]. This is because, if the point was not at the common tangent point, by keeping one of the terms fixed it is possible to move along the corresponding level set in a direction which decreases the other term thereby decreasing the cost as a whole as shown in Fig. 2.2a. In Eq. (2.58) the level set corresponding to the ℓ_1 term is a rotated square with corners at the axes and the level set of the quadratic term is a circle for the case of two dimensional data. In this case the chance of the circle intersecting the square is high at the corners rather than the lines especially when the relative size of the circle is high compared to that of the square which can be inferred from Fig. 2.2b.

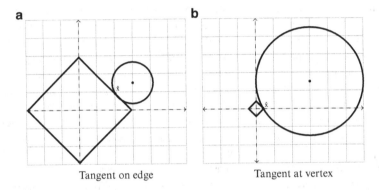

Tangent on edge   Tangent at vertex

Fig. 2.2 Illustration of sparsity inducing nature of ℓ_1 norm. (**a**) The solution point is a tangent point of the two level sets (of ℓ_1 and ℓ_2 norms). (**b**) As the size of the ℓ_2 term increases relative to the ℓ_1 term, the tangent point tends to be the corner of the ℓ_1 level set

Since an image is sparse only in the transform domain or the derivative domain, sparsity is used as a regularizer for estimation in these domains. In Chap. 7 we look at the blind deconvolution problem using sparsity inducing regularizers in the derivative domain and in the wavelet domain. Using ℓ_1 or a variant of it as the regularization term leads to a cost function which is the sum of a differentiable data term and a non-smooth regularizer term. Since the cost function is non-differentiable in this case gradient descent methods will not work. One method of arriving at the optimum of such cost functions is the iterative shrinkage thresholding algorithm (ISTA) which is explained in the next section.

2.6 Optimization Techniques

In this section we look at two of the approaches we use in the monograph for arriving at an optimal solution. One is the alternate minimization technique and the other is iterative shrinkage thresholding algorithm.

2.6.1 Alternate Minimization

As seen in Sect. 2.3.4, blind deconvolution is a bilinear ill-posed problem. Since both the image and the PSF are unknowns, regularizers are used for both. Using square of ℓ_2 norm of the error $(\underline{y} - K\underline{x})$ as the data term the cost function becomes:

$$C(\underline{x}, \underline{k}) = \| \underline{y} - K\underline{x} \|^2 + \lambda_x R_x(x) + \lambda_k R_k(k),   \qquad (2.59)$$

where λ_x and λ_k are image and PSF regularization factors, and $R_x(x)$ and $R_k(k)$ are the respective regularizers. Since the problem is bilinear in nature, keeping one of the variables fixed makes the data term a quadratic function. If the functions $R_x(x)$ and $R_k(k)$ are chosen to be quadratic functions, keeping one of the variables fixed makes the whole cost quadratic. This makes alternate minimization a natural choice for minimizing the cost given in Eq. (2.59). In addition, keeping one of the variables constant and minimizing w.r.t. the other variable makes the problem a linear ill-posed one at each iteration. The alternate minimization algorithm proceeds as follows:

1. Start with an initial value of $\underline{k}^0 = \underline{k}_{initial}$ and $\underline{x}^0 = \underline{y}$.
2. At the ith iteration, keep $\underline{k}$ fixed at $\underline{k} = \hat{\underline{k}}^{i-1}$ and minimize Eq. (2.59) to obtain a new estimate of $\underline{x}$, $\hat{\underline{x}}^i$. Here the initial value of $\underline{x}$ is $\hat{\underline{x}}^{i-1}$.
3. Keep $\underline{x}$ fixed at $\hat{\underline{x}}^i$ and minimize the cost to obtain $\underline{k}$ update as $\hat{\underline{k}}^i$, with initial condition on $\underline{k}$ as $\hat{\underline{k}}^{i-1}$.
4. Repeat steps 2 and 3, till some stopping criterion is reached.

The initial value of $\underline{k}$, $\underline{k}_{initial}$, is chosen as the 2D-discrete impulse. In this monograph we use alternate minimization for solving the blind deconvolution problem. At each step of the alternate minimization method, depending on the cost function obtained, a gradient based method is used to arrive at the optimal points of each step. The AM algorithm generates a sequence of image and PSF pair – $\{(\hat{\underline{x}}^i, \hat{\underline{k}}^i)\}$. We shall study the convergence of this sequence for the TV regularizer and the smoothness based regularizer in subsequent chapters.

2.6.2 Iterative Shrinkage/Thresholding Algorithm

We observed in the previous section that sparsity based solutions use ℓ_1 norm or its variants as a regularizer. With ℓ_1 norm as the regularizer, the cost function becomes the sum of a quadratic data term and a non-differentiable term coming from the regularizer. The usage of ℓ_1 norm calls for optimization methods which can handle non-differentiable functions. Optimization of such cost functions, for function of a single variable is given below. Let the general form of the cost function be

$$C(x) = f(x) + g(x),$$
(2.60)

where both $f(x)$ and $g(x)$ are convex functions with $f(x)$ being a smooth quadratic data term and $g(x)$ a non-smooth regularizer. For solving problems of the type in Eq. (2.60), Moreau's [8, 100] proximal map can be used. Given a closed convex function $h(x)$, the proximal map associated with $h(x)$ is

$$\text{prox}_t(h)(x) := \underset{u}{\arg\min} \left\{ h(u) + \frac{1}{2t} \parallel u - x \parallel^2 \right\},$$
(2.61)

where t is a positive constant. Since $g(x)$ in Eq. (2.60) is non-smooth, one can use the subdifferentials [134] to obtain a minimizer. A vector $x*$ is a minimizer of Eq. (2.60) *iff*

$$0 \in t\nabla f(x*) + t\partial g(x*),$$

$$0 \in t\nabla f(x*) + x * -x * +t\partial g(x*),$$

$$(I + t\partial g)x* \in (I - t\nabla f)x*,$$

$$x* \in (I + t\partial g)^{-1}(I - t\nabla f)x * .$$
(2.62)

In this set of equations $\partial g(x)$ is the subdifferential of the closed convex non-smooth function $g(x)$. From the last step of Eq. (2.62), an iteration to obtain $x*$ can be written as

$$x^i = (I + t^i\partial g)^{-1}(I - t^i\nabla f)x^{i-1}.$$
(2.63)

It can be shown that [8]

$$(I + t^i\partial g)^{-1}(x) = \text{prox}_t(g)(x).$$
(2.64)

Using Eq. (2.64) in Eq. (2.63) and using the definition of proximal map given in Eq. (2.61),

$$x^i = \text{prox}_t(g)(x^i - t^i\nabla f(x^{i-1})),$$

$$= \underset{x}{\arg\min} \left\{ g(x) + \frac{1}{2t^i} \parallel x - (x^{k-1} - t^i\nabla f(x^{i-1})) \parallel^2 \right\}.$$
(2.65)

When $g(x) = \parallel x \parallel_1$, the ℓ_1 norm of x, Eq. (2.65) reduces to [8, 99]:

$$x^i = \mathcal{T}_{t^i}(x^{i-1} - t^i\nabla f(x^{i-1})),$$
(2.66)

where $\mathscr{T}_{t^i}$ is the shrinkage operator defined as

$$\mathscr{T}_\alpha(x^i) = (|x^i| - \alpha)_+ \mathrm{sgn}(x^i), \tag{2.67}$$

where $(a)_+ = \max(a, 0)$ and $\mathrm{sgn}(x)$ is the signum function. When $f(x)$ is the quadratic data term in Eq. (2.59) (with K constant), Eq. (2.66) becomes the iterative shrinkage threshold algorithm (ISTA) [29].

Chapter 3
Blind Deconvolution Methods: A Review

In this chapter we look at the different solution methods that have been proposed for blind deconvolution. Though the methods could be classified in many ways, in order to depict the evolution of the solution methods, we describe the different techniques under a broad classification of earlier and recent techniques, which is further split into deterministic and statistical techniques.

3.1 Earlier Approaches

Spectrum of image with motion blur caused by translational motion has spectral zeros and earlier approaches for estimating the PSF relied on these spectral zeros to estimate the blur and hence were applicable only for a limited type of blurs. We describe classical transform based methods in this section.

One of the earliest works reported in 1975 [153] uses homomorphic signal processing for deconvolution. Here in noise free case the reconstruction filter is the inverse filter and it is shown that in the noisy case it becomes the geometric mean of the inverse filter and the Wiener filter. To estimate the inverse filter multiple images blurred by the same PSF is required. Since this is not available the estimation is done by splitting the observed image into smaller images with size much larger than the PSF size to avoid edge effects. The log spectrum of these smaller images are used to estimate the PSF [153]. It is important that the phase information is known. Hence this method works only for simple blurs. Besides, this method works only if an identical recording system with a flat frequency response is available. Since this method is too limited to be of any practical use, more sophisticated methods were developed. The classical iterative methods are described next.

One of the first iterative methods for blind deconvolution [83] uses the idea of zero sheets [84]. It is shown in [83] that for a noise free case the individual components of a composite image can be recovered without using any filtering

© Springer International Publishing Switzerland 2014
S. Chaudhuri et al., *Blind Image Deconvolution: Methods and Convergence*,
DOI 10.1007/978-3-319-10485-0_3

operation provided the dimension of the individual components is greater than unity. An image which is a convolution of M individual components ($M > 1$) is said to be composite if

1. Each component has a compact support (S)
2. The spectrum of each component is zero on a single continuous surface [152] called as the zero sheet and
3. The M zero sheets are distinct – they intersect only at discrete points in the 2S dimensional space.

The argument is that the spectra are necessarily zero on 2S-2 dimensional hyper-surfaces in a 2S dimensional space. In the absence of noise, the component zero sheets can easily be extracted from the zero sheet of the convolution. Knowing the zero sheet, the Fourier transform of the image is computed in an iterative manner, from which the original image is obtained using inverse Fourier transform, the details of which are given in [83]. This algorithm has the disadvantage of being highly sensitive to noise and is also computationally complex. The component zero sheets start coalescing in the presence of noise and hence the algorithm works only for very low levels of noise. Ghiglia et al. [42] gave an algorithmic framework for the idea introduced in [83]. Though the implementation was made robust, the complexity and the ability to handle noisy data was not improved in [42].

Another iterative approach to blind deconvolution was provided by Ayers et al. [5], using an approach which is quite different from that of [83]. Starting with an initial image, the original image and the PSF are estimated iteratively. The algorithm switches between the spatial and frequency domains. Inverse filtering is used to estimate the image from the PSF and vice versa. Image plane constraints, such as positivity, and any other information available regarding the PSF are used to reduce the error associated with inverse filtering. Though ill-posedness is not handled directly, at each step of the inverse filtering, a Wiener-like filter is used, which gives the algorithm a better robustness to noise compared to [83]. Ability of the algorithm to handle a wide range of noise is not studied, and the estimates depend on the initial guess. The computational complexity is less compared to [83].

Use of iterative blind deconvolution for handling images with speckle noise is reported in [103] and [158]. By combining the ideas of [83] and [5], Davey et al. [30] implemented an iterative blind deconvolution algorithm by considering a general image whose pixel values can be a complex number. But the algorithm is limited to deconvolving only complex valued images which contain simple structures.

It has been shown in Schafer et al. [140] that for the case of non-blind deconvolution the constraints of positivity and finite support for the image offers the possibility of restoring high frequency components lost due to blurring. This fact was used by Kundur et al. [78] to develop a blind deconvolution algorithm applicable to an image consisting of finite support objects against a uniformly black, gray or white background. The algorithm is recursive in nature and is named as the non-negativity and support constraints recursive inverse filtering (NAS-RIF). The algorithm uses a variable coefficient FIR filter to estimate the image. The estimated image is subjected to nonlinear constraints which project the estimated image into a space that represents the known characteristics of the original image. The error between the estimated image and the projected estimate is used to update the filter coefficients. The optimization procedure, which updates the filter coefficients, minimizes a cost function that is convex in nature and guarantees convergence to the global solution. Though uniqueness of the solution is not guaranteed, it is shown that for most practical images an unique solution is obtained. In the presence of noise, the convergence point is not necessarily the best estimate of the original and termination of iterations is determined by the visual quality of the estimate. Though the authors mention that regularization can be used for handling noise it was provided as a choice which the user can make and was not implemented.

Methods which provide better estimates and which can handle noise use more sophisticated techniques like regularization or statistical methods. These methods are described in the next section followed by a section on alternate solutions which are based on Radon transform, hardware support or multichannel image data.

3.2 Regularization Based Techniques

We discussed in Chap. 2 that blind deconvolution is a bilinear ill-posed problem and that regularizers are needed to handle the ill-posedness. Regularization techniques are well established in solving the deconvolution problem for a known blur – we use the terms blur, kernel and PSF interchangeably. We give a brief overview of regularization applied to non-blind deconvolution to bring out the difficulties faced when the same technique is applied for the blind deconvolution problem.

Deconvolution is a linear ill-posed problem as explained in Sect. 2.3.1. Linear ill-posed problems are well studied and as can be seen from the literature, there exists a mature theory for the deblurring problem complete with convergence analysis and rate of convergence estimates. We review a few of the early regularization based techniques in this section.

3.2.1 Constrained Least Squares

It was identified by Sondhi [147] that image restoration is an ill-posed problem by showing the equivalence of image restoration to the solution of Fredholm equation of the first kind. Phillips [124] used a smoothness measure to constrain the least square estimation of one dimensional signals. This method was extended to two dimensional signals by Hunt in [54].

In constrained least squares (CLS) [54,65,69,70,140], a constraint is imposed on the estimated parameter which is the sharp image in this case and the resulting cost function is minimized subject to a constraint on the noise variance. The estimated image $\hat{\underline{x}}$ is

$$\hat{\underline{x}} = \arg\min_{x} \| C \underline{x} \|, \tag{3.1}$$

subject to

$$(y - K\underline{x})'(y - K\underline{x}) = n'n, \tag{3.2}$$

where C is the matrix which enforces the constraint on the solution and the prime indicates the transpose. Since natural images are in general smooth, a frequently used constraint is smoothness of the estimated image. It is assumed that information concerning the noise variance is available which becomes the a priori information that regularizes the ill-posed problem. Comparing with Eq. (2.32) with $\underline{x}$ replaced by $\mathscr{L}\underline{x}$ in Sect. 2.3.2, it may be noted that this is equivalent to Tikhonov regularization. In [54], the image is estimated using Lagrangian minimization. The image is estimated in a computationally efficient manner by making use of the fact that the convolution matrices are block circulant and are diagonalized by the Fourier transform and the Lagrange multiplier is estimated in an iterative manner. Another type of regularization was proposed by Miller [101] which uses as regularizer a bound on the smoothness in addition to assuming that the noise variance is available. Here the solution is taken as the vector $\underline{x}$ which satisfies both the constraints

$$\| Cx \| \le E, \tag{3.3}$$

$$\| Kx - y \| \le \epsilon, \tag{3.4}$$

where E sets an upper bound on the deviation from smoothness of x, and ϵ is the upper bound on the noise variance. It is shown in [141] that a priori knowledge constraints the solution to certain sets which can be considered as ellipsoids. Specifically the a priori information can be specified in the form $x \in Q_x$, where Q_x is a set of signals with known properties and the corresponding ellipsoid is

$$Q_x = \{x \mid (x - c_x)' \Gamma^{-1}(x - c_x) \le 1\}, \tag{3.5}$$

where c_x is the center of the ellipsoid and the eigenvectors and eigenvalues of Γ decide the orientation and length of the axes of the ellipsoid Q_x. The conditions in Eqs. (3.3) and (3.4) define two sets which can also be seen as ellipsoids. The condition in Eq. (3.3) defines the ellipsoid

$$Q_x = \{x \mid \| Cx \|^2 \le E^2\}, \tag{3.6}$$

which has center $c_x = 0$ and $\Gamma^{-1} = C'C/E^2$. The second condition in Eq. (3.4) puts a bound on the noise variance. This confines the zero mean noise to an ellipsoid Q_n defined as

$$Q_n = \{n \mid n' \Gamma_n^{-1} n \le 1\}, \tag{3.7}$$

where $\Gamma_n = \epsilon^2 I$, I being the identity matrix. Given an observation y, the condition Eq. (3.4), thus restricts x to lie in a set $Q_{x|y}$ defined as,

$$Q_{x|y} = \{x \mid (y - Kx) \in Q_n\}. \tag{3.8}$$

Fig. 3.1 Representation of
Eqs. (3.3) and (3.4) as
ellipsoids

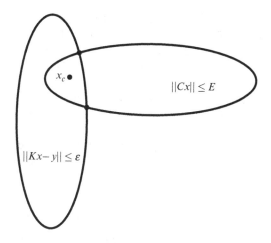

Since both the conditions need to be satisfied, the solution x should lie in the
intersection of the two sets ($Q_0 = Q_x \cap Q_{x|y}$). This is shown in Fig. 3.1. To
geometrically define the intersection of the two ellipsoids an ellipsoid which bounds
the intersection is defined. The center (x_c) of the bounding ellipsoid is taken as the
restored image, which is given by

$$(K'K + \alpha C'C)x_c = K'y. \tag{3.9}$$

A detailed analysis of both the methods, namely Hunt's [54, 101], is given in
[65]. An iterative method for estimating the image using the method of [101] is also
provided. Solution of both [54] and [101] reduces to solving the equation

$$(K'K + \alpha C'C)x = K'y, \tag{3.10}$$

where $\alpha = (\epsilon/E)^2$ for Miller's method (same as Eq. (3.9)) and is the Lagrange
multiplier in the case of CLS. The advantage of Miller's methods over CLS is that
the Lagrange multiplier need not be estimated in this case. The choice of C is guided
by the fact that the condition number of $K'K + \alpha C'C$ should be less than the
condition number of K i.e.

$$P(K'K + \alpha C'C) < P(K), \tag{3.11}$$

where $P(K) = \| K \|_2 \| K^\dagger \|_2$ which is a modified definition of condition number
of a matrix, here $K^\dagger$ is the Moore-Penrose generalized inverse. If one assumes that
$K'K$ and $C'C$ commute then the above condition is satisfied. This assumption is
satisfied when K and C are linear shift invariant systems. In this case the condition
number becomes

$$P(K'K + \alpha C'C) = \frac{\max_i (\mu_i^2 + \alpha\sigma_i^2)}{\min_i (\mu_i^2 + \alpha\sigma_i^2)}, \tag{3.12}$$

where μ_i's are the eigenvalues of K and σ_i's are the eigenvalues of C. From Eq. (3.12) it is seen that for Eq. (3.11) to be satisfied, the sequence σ_i should increase with i since μ_i is a decreasing sequence w.r.t. i. It may be noted that the eigenvalues μ_i and σ_i also correspond to the DFT of the filters corresponding to the blur K and the regularizer C. Hence the requirement on σ_i being a monotonically non-decreasing sequence corresponds to choosing C as a high pass filter. For completeness sake we give a brief overview of the convergence analysis given in [65, 66].

To estimate x, Eq. (3.10) needs to be solved. This can be done iteratively instead of using the costly matrix inversion, with the steps in the iteration being

$$x_0 = \beta K' y,$$

$$x_{i+1} = \tilde{x}_i + \beta(K' y - (K' K + \alpha C' C))\tilde{x}_i \tag{3.13}$$

$$= (I - \beta(K' K + \alpha C' C))\tilde{x}_i + \beta K' y, \tag{3.14}$$

where $\tilde{x}_i = P x_i$, P being the projection operator which represents hard constraints like the range of values taken by the variable, and is a way of incorporating additional prior information, and β is a parameter which ensures convergence and is selected so as to maximize the rate of convergence. The signals which satisfy hard constraints define convex sets and the applying a hard constraint is equivalent to projecting onto a convex set [177]. In [177] it is also shown that projection operators onto closed convex sets are non-expansive i.e., $\| Px - Py \| \leq \| x - y \|$. The condition for the iterations in Eq. (3.14) to converge is

$$\| (I - \beta(K' K + \alpha C' C))(P x_i - P x_{i+1}) \| \leq \eta \| x_i - x_{i+1} \|, \tag{3.15}$$

for $0 \leq \eta < 1$. Since P is a non-expansive operator, the condition in Eq. (3.15) is satisfied if the operator $I - \beta(K' K + \alpha C' C)$ is contractive. The fixed point of the iteration in Eq. (3.14) is x_c if x_c belongs to the Q_p which is the intersection of the convex sets corresponding to the hard constraints imposed by the projection operators P, if multiple of them are used. It $x_c \notin Q_p$ then the fixed point is a vector which belongs to Q_p and is closest to x_c. In general $K' K + \alpha C' C$ is singular and hence $I - \beta(K' K + \alpha C' C)$ is non-expansive, which means that multiple number of fixed points can exist. In this case the iteration converges to the minimum norm solution of Eq. (3.10) plus the projection of x_0 onto the null space of $(K' K + \alpha C' C)$ provided that $0 < \beta < 2 \| K' K + \alpha C' C \|^{-1}$, and $K' y$ is in the range of $(K' K + \alpha C' C)$. If the matrices $K' K$ and $C' C$ commute, C can be chosen so that $K' y$ is in the range of $(K' K + \alpha C' C)$. The work in [65] has also proposed a modified algorithm which incorporated the properties of human visual system [3] yielding restored images that are visually better. Appropriate selection of the regularization parameter for such a deconvolution has been addressed in several works [41, 63, 132, 155]. The requirement of commutativity of $D' D$ and $C' C$ restricts the restoration systems to be linear shift invariant. However use of non-linear regularizers leads to better results [17, 39, 115]. Since non-blind deconvolution is a linear ill-posed problem it is well analyzed and convergence can be proved unlike in the case of blind deconvolution.

In blind deconvolution in addition to lack of stability one has to address the problem of uniqueness of solution also, since there could be many (x, k) pairs which satisfy the image formation model Eq. (1.9). Since the image and the PSF are both unknowns, regularizers are used for both. Though positivity and size constraints [30,82,97,169] are useful in reducing the number of admissible solutions in blind deconvolution, they are by no means enough. Smoothness constraint of the image has been implicitly incorporated into the ML-based algorithms via the space-invariant AR image model [81,154] and into the generalized cross-validation (GCV) algorithm via the space-invariant regularization [133]. But it has been shown in [79] that space-invariant restoration methods induce ringing artifacts around sharp intensity transitions. It was demonstrated in [79] for the non-blind case that space-adaptive restoration reduces ringing artifacts. Incorporation of space-variant models in ML and GCV based algorithms complicates these estimation methods. You et al. [171] extended the work in [79] to include space-adaptive regularization for the blind restoration problem.

3.2.1.1 Quadratic Regularizers

You and Kaveh [171] used a space-adaptive quadratic regularizer for both the image and the PSF. The regularizer used was the Laplacian and space-adaptivity was introduced by using weights for the data term and the two regularizers. Usage of space adaptivity exploits the piece-wise smooth nature of the image which leads to a better solution than the case of ML/GCV which uses only the smoothness of the image. The cost function solved in this case is

$$C(x, k) = \| W_1(\underline{y} - K\underline{x}) \|^2 + \lambda_x \underline{x}' D' W_2 D\underline{x} + \lambda_k \underline{k}' D' W_3 D\underline{k}, \qquad (3.16)$$

where the W's are diagonal matrices which introduce the spatial-adaptiveness, λ_x and λ_k are the image and the PSF regularization factors. Alternate minimization (AM) is used to estimate the image and the PSF iteratively. The regularizer used here is a smoothness inducing one, and the regularization parameters control the trade-off between fidelity to observation and smoothness of the restored image. Approximate estimates of both the regularization factors is also arrived at. But the problem with this approach is over smoothness of the estimated image, and the possibility of the solution being a local minimum. The results showed a marked improvement compared to the ML/GCV based approaches to blind deconvolution. A convergence analysis of this algorithm is given in [17], which is discussed in Chap. 5.

In [96] using the local variance of the degraded image, upper and lower bounds on pixel values are derived, which is used to overcome the problem of convergence to local minima in [171]. This significantly improved the accuracy of the image and PSF estimates. Molina et al. [105] modified the approach in [171] by using Dirichlet distribution to model the blur. Steepest descent was used for reaching the minima, but this method also suffers from the problem of solution being a local minima.

The algorithm in [171] is extended to handle shift-variant blurs in [172, 175]. Shift-variant blurs occur in the case of motion blur when objects move differently against a background and in the case of out-of-focus blur when the scene has a depth variation. The PSF and image regularizers are chosen such that the requirements of weighted smoothing (shift-adaptive regularization) of the image, and prevention of smoothing across the edges are both met. These two conditions are achieved using anisotropic diffusion (see the references in [172] for details). The paper considers only blurs which can be parametrized, the results of blind restoration done for a horizontal motion blur on a synthetic image is provided. This method also suffers from the problems of local minima and non-uniqueness of solution and is not suitable for highly textured images, which can cause difficulties in anisotropic diffusion based smoothing [175].

3.2.2 Total Variation as a Regularizer

Following the work in [79], You et al. [174] used anisotropic diffusion as a regularizer to provide an orientation selective regularization for the restoration problem. This is achieved by selecting the diffusion constant such that in smooth areas where the magnitude of the gradient is below a threshold, isotropic smoothing takes place else an edge preserving diffusion takes place which smoothes tangential to the edge. This way, small variations in image intensity due to noise are smoothed out and the edges are preserved. Anisotropic regularization was extended to blind deconvolution in [173, 176] to regularize both the image and PSF estimation. The cost function used is [176]

$$C(x,k) = \frac{1}{2}\int_{\Omega} e^2(u,v)dudv + \lambda \int_{\Omega} \kappa(|\nabla x(u,v)|)dudv + \gamma \int_{D} \beta(|\nabla k(u,v)|)dudv,$$
$$(3.17)$$

subject to the constraints

$$k(u,v) \geq 0, \quad (u,v) \in D,$$

$$\sum_{(u,v)\in D} k(u,v) = 1.$$

In Eq. (3.17), Ω and D correspond, respectively, to the region in space where the image and PSF are defined. The function $\kappa(.)$ is an increasing function chosen such that the minimization of this term leads to a diffusion operation which progresses such that image is smoothed only in the direction of the edges. The function $\beta(.)$ plays a similar role for the PSF. The term $e(x, y)$ is the data fidelity term which gives the difference between the observed image and the estimated image. The unknowns are estimated using alternate minimization. This method also suffers from the problem of local minima and non-uniqueness of solution. In addition highly textured images can cause difficulties owing to the anisotropic diffusion. It is also shown that the effect of this regularizer is equivalent to that obtained by minimizing the total variation by selecting $\kappa(.)$, $\beta(.)$ as $\kappa(s) = s$, $\beta(s) = s$.

With choice of $\kappa(.)$ and $\beta(.)$ as mentioned above Chan et al. [16] proposed a fast alternate minimization based algorithm for estimating the image and the PSF. The cost function in Eq. (3.17) gets modified as

$$C(x,k) = \frac{1}{2} \int_{\Omega} (k * x - y)^2 du dv + \lambda \int_{\Omega} |\nabla x| du dv + \gamma \int_{D} |\nabla k| du dv. \qquad (3.18)$$

Positivity constraints are used for both image and PSF. Additional constraints of image mean preservation and PSF symmetry are used to reduce the size of the solution space. An improved performance over [171] was observed due to the edge preserving nature of the TV regularizer. Each step of the AM algorithm gives rise to a PDE which was solved using the lagged diffusivity fixed point method [161], wherein the denominator for the derivative of the TV term is lagged by one step of the iteration. The results demonstrate that even if the PSF is not recovered exactly, the error in the recovered image is not very high. It is also demonstrated that the technique is good for identifying PSF's without edge like the Gaussian blur, but the convergence rate of the AM is slower in this case. Use of total variation as a regularizer is reported by many other authors also, a few of which we have mentioned below.

Rodríguez et al. [135] propose an iteratively reweighted norm (IRN) algorithm which minimizes the generalized TV function efficiently. A summary of the existing methods for minimizing the TV function is also given in [135]. The algorithm is tested for denoising and deblurring only. Though the blind case was not considered, it could be used for blind deconvolution also.

Bar et al. [7] observed that, for the TV regularizer based method of Chan et al. [16], there is an excessive dependence of the recovered kernel on the image characteristics. This behavior is attributed to the fact that the algorithm imposes only mild constraints on the shape of the reconstructed kernel, which allows the distribution of edge directions in the image to have an influence on the shape of the recovered kernel via the regularization factors. By assuming a parametric form for the blur and by using the fact that blind deconvolution and image segmentation are tightly coupled problems, authors in [7] propose a semi-blind image deconvolution which iteratively alternates between segmentation, parametric blur identification and restoration. The cost function consists of a quadratic data fitting term and an image regularizer which converges to the Mumford-Shah functional [109], and a smoothness regularizer for the blur. The risk of convergence to local minima inherent in most of the blind deconvolution algorithms is present in this case also. It may be noted that this method does not perform well in presence of noise since the edge map estimation part of segmentation breaks down in presence of noise. Methods to stabilize the edge map estimation, leading to reduction in restoration artifacts due to noise was proposed in [182].

A regularized version of the NAS-RIF [78] which was mentioned in Sect. 3.1 is proposed in [113]. A regularization is applied for the FIR inverse filter used in NAS-RIF so as to reduce noise amplification. At each iteration the eigenvalues of the convolution matrix corresponding to the FIR filter, which are below a threshold are suppressed and the resulting error is also made a part of the overall cost function. The authors have mentioned that a total variation based image regularization could be used to estimate the image, but was not implemented.

In [107], the result of blind deconvolution using TV was drastically improved by using pre-processed reference image obtained via the shock filter as the initial condition. Shock filter [118] reverses the effect of applying a Gaussian filter and creates a discontinuity at edge locations thereby providing information on the location of edges. As the authors have observed, this may not represent the true image when the blur kernel is not Gaussian. Using TV as regularizer gives robust reconstruction of the image and the PSF. A comparison of using ℓ_1 and ℓ_2 norm for the data fitting term is also given. It was shown experimentally that using ℓ_1 norm for the data fitting term leads to better results while estimating the image and that ℓ_2 norm for the data fitting term is a better choice for estimating piece-wise constant kernels, i.e., for each iteration of the AM algorithm, two different norms are used for the data fitting term depending on whether the image is being estimated or the kernel.

Liao et al. [91] proposed a blind deconvolution scheme in which the regularization parameters were estimated using GCV. Variational Bayesian methods tend to give a better result compared to total variation and quadratic regularizers since the regularization parameters are also variables to be estimated in the optimization procedure. In [91] by using generalized cross validation (GCV) for estimating the parameters, better results are obtained compared to conventional regularization based approaches. TV is chosen as the image regularizer and the PSF regularizer is the square of the Euclidean norm of the second-order difference among the pixels. AM is used to estimate the image and the PSF. For this formulation AM can be seen as alternate steps of denoising and deblurring. The denoising step is implemented using the methods proposed in [51, 164], which are fast implementation of TV regularizers. These methods use either an auxiliary variable along with a quadratic term, or a variable splitting and quadratic penalty method, where auxiliary variables replace the gradients in the nondifferentiable TV term and quadratic penalty term is used to make the auxiliary variable close to the gradient. GCV is used to estimate the regularization parameter. Finding the regularization parameter for PSF using GCV is direct since only quadratic functions are involved. For the image case, since variable splitting is used, an additional AM is involved in the image estimation step which involves minimizing a quadratic function of the image. The image regularization parameter is obtained by applying GCV at this step. The results obtained are competitive with the variational Bayesian methods with Student's-t prior or a total variation prior.

3.2.3 Other Regularizers

A regularization based blind image deblurring scheme that makes only a weak assumption of limited support about the blurring filter is proposed in [1] and is capable of handling different types of blurring degradations. A novel image prior which depends on the image edges is used. This prior favors solutions with sparse edges since natural images have sparse edges. A deblurred solution is achieved by first concentrating on the main edges and progressively dealing with weaker edges. The regularization parameter is decreased in each iteration. A sparsity controlling parameter is also used which is also decreased in each iteration. A set of edge detection filters are used for forming the image prior. An AM scheme is used to update the image and the PSF. Any additional information available about the PSF can be used as the PSF regularizer. Results are shown for the single frame, multi-frame and color image cases. Experimental tests show good results, though there is a flattening effect in the image.

Another iterative regularization based scheme which takes into consideration the difference between the image and the PSF spaces is given in [21]. In addition to using spatially adaptive smoothness regularization, a reinforcement learning term is also used to incorporate the parametric structure of blur. Alternate minimization is used for estimating the unknowns, along with additional operations in the blur estimation step for integrating information regarding structures exhibited by the blur. Results are shown only for synthetic images of small size.

Krishnan et al. [76] proposed a sparsity based regularizer for images. In [2, 36] it was observed that joint estimation using MAP gives only trivial results and the reason was attributed to the nature of the image prior [2] which is usually a function of the gradients of the image. Such an image prior has lower cost for a blurred image than for a sharp image leading to choice of the blurred one as the minimum in the trivial case. One way to overcome this is by using an image prior which has a cost that increases with the blur amount. Such a prior was proposed in [76], which is the ratio of the ℓ_1 norm to the ℓ_2 norm of the derivative of the image. It is claimed that ℓ_1/ℓ_2 satisfies the heuristic criteria mentioned in [55] for being a sparsity measure. The ℓ_1 norm part prefers a sparse solution and the ℓ_2 denominator normalizes the measure. Since ℓ_1/ℓ_2 norm favors sparse solution, the deconvolution is done in the sparse derivative domain. Only motion blur was considered and ℓ_1 norm was used as regularizer for the PSF since motion blur is sparse in nature. AM is used to solve the cost function which is formulated in the derivative domain to obtain the estimate of image derivative and PSF. But the derivative domain image estimate is discarded and the estimated PSF is used to find the image using a non-blind procedure, which uses hyper-Laplacian priors to model the image gradient distribution [75]. One of the steps in the alternate minimization procedure which is non-convex is solved using a lookup table which speeds up the deconvolution algorithm.

3.3   Statistical Methods

In this section we give an overview of solutions to the blind deconvolution problem based on ML, MAP, variational methods and higher order statistics.

3.3.1   Maximum Likelihood Restoration

In classical approaches ML has been used to estimate the model parameters of an ARMA model in which the image is modeled as an auto regressive process(AR) and the PSF is modeled as a moving average (MA) process. This gives

$$x = Ax + w$$
$$y = Bx + v, \tag{3.19}$$

where x, y are the original and observed images, respectively, A the matrix of AR model parameters, w and v are independent zero mean AWGN with variances σ_w^2 and σ_v^2, respectively. B forms the convolution matrix containing the MA parameters. In blind deconvolution A, B, x, and the noise variances are unknowns. Several methods for estimating the parameters using maximum likelihood (ML) and its variants and expectation maximization (EM) have been proposed in the literature which are briefly reviewed below.

In [154] Tekalp et al. have developed a spatial domain procedure which simultaneously identifies the PSF and the image model parameters. Conditional ML estimates of the parameters are derived for the noiseless and noisy observation cases. The blur PSF is modeled as a noncausal PSF, making recursive realization impossible. Hence the unknown PSF is decomposed into four quarter-plane convolutional factors, each being stable in its direction of recursion. These factors are identified and convolved to estimate the unknown PSF. In the noiseless case ML is used to estimate the parameters. In the presence of noise, the MA parameters cannot be uniquely determined, and the authors have used the method of Kashyap et al. [64] which uses correlation along with the mean conservation criterion (sum of PSF elements is unity) for estimating the MA parameters. The estimated parameters are used for the design of an efficient Kalman filter (reduced update Kalman filter [167]) for restoring the noisy blurred images.

Biemond et al. [11] have described a parallel identification and restoration procedure for images with symmetric, noncausal blurs. A parallel bank of Kalman filters is derived for doing the deconvolution. The ARMA model in Eq. (3.19) is rewritten by considering each row of the image as the variable instead of considering the pixels as the variables. This leads to a model which has a set of filters in parallel. The parameter estimation is done using the frequency domain representation of Eq. (3.19) incorporating the modification mentioned. Estimation process is split into an initial ARMA model estimation of a set of 1-D complex ARMA models in the column direction followed by another ARMA parameter estimation in the row direction. The ARMA model is expressed as an equivalent infinite order AR model and a linear estimation procedure [45] which yields a minimum phase solution, is applied. From this minimum phase solution, under the condition of blur symmetry, the noncausal MA blur parameters are reconstructed.

Lagendijk et al. [81] formulated the blur identification process as a constrained ML problem for noise-free, blurred images. An ARMA model is assumed with AR model for image and MA model for the blur. The model parameters are estimated using ML with the following constraints.

- The PSF of the blurring system has some symmetry or in general a number of PSF coefficients have identical values.
- Some symmetry is enforced on the image model in the sense that there are image model coefficients with identical values.
- The sum of PSF elements is unity.
- The AR model coefficients sum to unity.

These constraints impose linear relations between the image and blur model coefficients, which are used during the estimation process. This formulation has the disadvantage that there are local extrema to which the optimization can converge. A gradient based optimization method where the gradient of the likelihood function is calculated analytically is used. A similar work in which the gradient is calculated numerically is reported in [80]. In [80], the expectation maximization (EM) algorithm is used for ML blur identification. Here the blur parameters and the sharp image are estimated in an iterative manner, by solving linear equations only. But the performance becomes unsatisfactory for large blur sizes. In [68], an EM based method where the estimation is done in the frequency domain was proposed. It used the fact that when the image formation process is linear with Gaussian distributed noise [34, 67] the conditional density function is Gaussian leading to an easy evaluation of E-step and a straight forward minimization of the conditional density function.

3.3.2 Maximum A Posteriori Probability Restoration

As seen in Sect. 2.4, for using MAP to estimate the image and the PSF, the prior for the image and the PSF needs to be known. It was also pointed out that MAP estimation is similar to the regularized estimation with the priors acting as the regularizers. In this section we provide a brief overview of commonly used image priors and PSF models – we will later see the occurrence of these priors in variational Bayesian and regularization methods. Commonly used PSF models like the linear motion blur, atmospheric turbulence blur, and out-of-focus blur are described in [123]. The nature of the image like smoothness, texture, piecewise smoothness, etc. can be captured by specifying the probabilistic relations between the neighboring pixels or their derivatives, the same is applicable for PSFs also.

Essentially the image x, PSF k and the noise n are considered as samples of random fields, the distributions of which depend on parameters (Ω) called as the hyper-parameters [123].

The priors are described by a general exponential form [123] as explained in Chap. 2 while discussing the Gibbs distribution

$$f(x|\Omega) = \frac{1}{Z_x(\Omega)} \exp[-U_x(x, \Omega)], \tag{3.20}$$

$$f(k|\Omega) = \frac{1}{Z_k(\Omega)} \exp[-U_k(k, \Omega)],$$

where Z_x, Z_k are the normalizing terms known as the partition function. $U(.)$ is termed the energy function. Depending on the choice of $U(.)$ we get different priors. The total variation and anisotropic diffusion based regularizers can be written as priors using an appropriate function $U(.)$. For Gaussian models $U_x \propto \| Lx \|^2$ [123]. These models are termed as simultaneous or conditional autoregression models (SAR/CAR), depending upon the spatial dependence assumed. For a CAR model the spatial-dependence matrix is symmetric which need not be the case for SAR [145]. A commonly used L is the 2D-discrete Laplacian operator which constraints the derivative of the image [92,106]. In [68,80], the observed image was modeled as an ARMA process leading to an energy function of the form $U_x = \frac{1}{2} \| I - Ax \|^2_{\Lambda_v}$, where A is obtained from the AR filter coefficients and Λ_v is the covariance of the excitation noise.

Another way of defining the prior is by using equivalence of Markov random fields to Gibbs distribution [23, 181]. Here the energy function is defined in terms of clique potentials [90, 123] as $U = \sum_c V_c(x)$ where $V_c(x)$ is the potential function defined over the cliques. Using quadratic potentials one gets the Gaussian Markov random field (GMRF) or CAR. Isotropic diffusion, which is equivalent to a quadratic regularizer when discretized, and total variation induce probability densities of the form Eqs. (3.21) and (3.22), respectively [123].

$$f(x) \propto \exp[-\alpha \sum_i (\nabla_i^h f)^2 + (\nabla_i^v f)^2], \tag{3.21}$$

$$f(x) \propto \exp\left[-\alpha \sum_i \sqrt{(\nabla_i^h f)^2 + (\nabla_i^v f)^2}\right]. \tag{3.22}$$

Images being non-Gaussian ([165] and the references therein) has led to alternate image priors. It is observed that the gradient of the image has a sparse distribution or a heavy tailed one [2,36,165]. Mixture of a narrow and broad Laplacians each of which is centered around zero was used as prior in [87] to model the distribution of image gradient. In [86], which discusses a problem that is closely related to blind deconvolution, the fact that images have a sparse derivative was used in the reconstruction phase. Alternate priors were used by several authors which have been briefly reviewed in [76].

3.3.3 *Variational Methods*

Variational method was introduced as an alternative to alternate minimization and Markov chain Monte Carlo (MCMC) techniques to solve the Bayesian blind deconvolution problem. MCMC techniques are computationally intensive and it is claimed in [92] that the alternate minimization approach of solving the Bayesian blind deconvolution problem is suboptimal. Variational method is a generalization of the expectation maximization (EM) method [92]. Use of EM algorithm becomes difficult when it is not possible to specify the conditional PDF of the hidden variables, given the observation which is required in the E-step. In variational method an arbitrary PDF of hidden variables (here the image and the PSF) can be used. Let s represent the hidden variables and $q(s)$ the arbitrary PDF, a lower bound on the likelihood is obtained as [92]

$$F(q, \theta) = L(\theta) - KL(q(s) \parallel p(s|x)) = E_q(\log p(x, s)) + H_q, \qquad (3.23)$$

where x and s, are the observed and hidden variables, θ is a vector of the model parameters to be estimated, $L(\theta)$ is the likelihood, $KL(.)$ is the Kullback-Leibler divergence [77], $p(s|x)$ is the unknown conditional PDF of the hidden variables, E_q, the expectation w.r.t. $q(s)$ and H_q is the entropy of $q(s)$. $F(q, \theta)$ is a lower bound for $L(\theta)$. The variational method works by maximizing $F(q, \theta)$, and the location of maxima of $F(q, \theta)$ corresponds to the maxima of $L(\theta)$. This gives the variational EM steps as

$$\text{E-step: } q^{t+1} = \arg \max_q F(q, \theta^t),$$

$$\text{M-step: } \theta^{t+1} = \arg \max_\theta F(q^{t+1}, \theta). \qquad (3.24)$$

Assuming Gaussian probability density functions (PDF), the PDFs of the original image (x), PSF (k) and the noise (n) becomes, respectively

$$f(x) = N(\mu_x, \Sigma_x), f(k) = N(\mu_k, \Sigma_k), \text{ and } f(n) = N(0, \Sigma_n),$$

where μ represents the mean and Σ the covariance matrix. The unknown parameter vector becomes $\theta = [\mu_x, \Sigma_x, \mu_k, \Sigma_k, \Sigma_n]'$. In the blind deconvolution problem the hidden variables (s) are the original image (x) and the PSF (k), and the posterior distribution $p(s|\theta) = f(x, k|y, \theta)$ is unknown. Applying variational method, the posterior distribution $f(x, k|y, \theta)$ is approximated by the distribution

$$f(x, k; \theta) = f(x; \theta) f(k; \theta).$$

The unknown parameters are estimated by following the steps in Eq. (3.24).

A Gaussian prior was used [92] for the image and a simultaneous autoregressive (SAR) prior was used for the PSF. In the E-step the parameters of the arbitrary PDF is updated and three different methods are provided for implementing the M-step, the details of which are available in [92]. The results of blind deconvolution exhibits non-negligible amount of ringing. A similar method termed the ensemble learning was proposed in [102] for blind deconvolution. But this method does not make use of the pixel correlation which exists in natural images. Usability of variational method for blind deconvolution was demonstrated in [180].

Molina et al. [106] uses simultaneous autoregressions as prior distributions for the image and blur, and Gamma distributions for the unknown parameters of the priors and the image formation noise. Variational methods were used to approximate posterior probability of the unknowns. Though it compared well with other existing algorithms of that time, the result shows artifacts. This approach fails to model the edges in the image or PSF and also fails to estimate the PSF support.

As mentioned earlier, Bayesian formulation which introduces prior models on the image, blur and their model parameters, imposes constraints on the estimates, that act as regularizers. In [6], a variational method that uses the TV function as the image prior and the simultaneous autoregressive model as the blur prior was proposed. A Gamma distribution is used to model the hyper-parameters for the image, blur PSF and the image formation noise. Experimental results demonstrate high quality restoration for both synthetic and natural images.

A Bayesian model which permits reconstruction of image edges, finds the PSF support and takes into consideration the spatial PSF correlations was reported by Tzikas et al. [159]. In [159] the PSF is modeled as a linear combination of kernel functions located at all pixels of the image. The weights of the kernel is assumed to have a heavy tailed Student's-t distribution, which favors sparse models. This forces most of the kernel weights to zero and limits the support of the PSF. The first order local differences (horizontal and vertical) in the image is modeled using a Student's-t PDF. A two-level hierarchical model for noise is used with the prior being Student's-t (Gaussian noise with hyper-parameters assumed to be Gamma distributed). An approximate posterior distribution is obtained using variational methods, which reduces to minimizing the KL divergence between the original and modeled distribution. The results are superior to that of the Gaussian based variational method [92] and TV regularizer based method [16] for the case of small sized PSFs at low noise levels. For large sized PSFs and high noise levels, ringing artifacts are observed.

3.3.4 Higher-Order Statistics

Higher-order statistics (HOS) have also been used in different ways for blind deconvolution. The higher-order statistics of a stationary process are generally referred to as cumulants and their spectra as polyspectra [48, 98]. The third-order measures relate to the signal skewness and the fourth-order to the signal kurtosis

[114]. HOS is used when there is deviation from Gaussianity and also to detect and characterize nonlinearities in signals or for identifying nonlinear systems [98, 114].

Higher order spectral methods were used in blind deconvolution of ultrasound images in [160]. Here the tissue response is non-Gaussian and HOS methods are adopted for estimating a non-parametric form of the PSF using the third order cumulant and its Fourier transform – the bispectrum.

A Bussgang process [48, 56] has its autocorrelation function equal to the cross-correlation between the process and the output of a zero-memory nonlinear function of the process. HOS is implicit in Bussgang process [48, 56] because of the non-linearity and Bussgang process has been used extensively for one-dimensional blind deconvolution. In [121], the Bussgang blind equalization algorithm for the multichannel case has been extended to multichannel blind image deconvolution. This is an iterative method which gives a multichannel Wiener filter followed by a nonlinear estimator. Suitable nonlinear functions for non-Gaussian spatially uncorrelated and correlated images are derived in [121]. HOS has also been used for blind deconvolution of bi-level images with unknown intensity values [59] and also for deconvolving images obtained from a coded aperture camera, where the camera and scene are static [95].

3.4 Motion Deblurring

With the proliferation of consumer photography, research on motion blur removal is receiving considerable attention as is evident from the number of publications in this area in the recent past [2, 36, 60, 144]. Hence we are providing a separate section for discussing motion deblur using blind deconvolution, though most methods used are similar to those discussed in the earlier sections. A seminal paper in this area following which many other works were reported, is the one by Fergus et al. [36]. We start with a discussion of this followed by certain key papers (not an exhaustive list) in this area.

Fergus et al. [36] have approached the blur removal problem by using a prior which describes natural images in a better manner than the priors mentioned in the above sections. An additional improvement was their use of Bayesian approach which allowed them to estimate the blur kernel implied by a distribution of probable images. It was shown by Field [37] that real world images exhibit heavy tailed distributions in their gradients. This distribution of gradient magnitudes is represented using a zero mean Gaussian mixture model, the parameters of which are estimated from a natural image using the expectation maximization (EM) algorithm. In addition to the blurred image, the user has to input a rectangular patch in the blurred image, an upper bound on the number of pixels in the blur kernel and an initial guess about the orientation of the blur kernel. Since motion blur is sparse in nature and is always positive, the kernel prior was taken as a mixture of exponentials, the parameters for which were estimated from a set of kernels estimated from low-noise real images. The failure of MAP to provide a non-trivial solution was first reported in this paper [36]. The reason for this was attributed to the tendency of MAP to minimize all the image gradients in a uniform manner, whereas a natural image does have some strong gradients. Since MAP failed to produce non-trivial solutions, a variational approach was used to estimate the blur kernel.

In [36], since distribution of image gradients is used as the prior, the image derivatives are used in the estimation step instead of pixel values. This gives the a posteriori probability for the image gradient and kernel as

$$p(\nabla x_p, k \mid \nabla y_p) \propto p(\nabla y_p \mid \nabla x_p, k) p(\nabla x_p) p(k),$$

$$= \prod_i N(\nabla y_p(i) \mid k \circledast \nabla x_p(i), \sigma^2) \prod_i \sum_{c=1}^{C} \pi_c N(\nabla y_p(i) \mid 0, v_c) \prod_i \sum_{d=1}^{D} \pi_d E(k_j \mid \lambda_d),$$

$$\tag{3.25}$$

where x_p and y_p are the original and blurred image patches, ∇ represents the gradient, and k, the PSF. The image prior is a mixture of C zero mean Gaussians, with variance v_c and a weight of π_c for the cth Gaussian. The PSF prior is taken as a mixture of D exponentials (E) with parameter λ_d and weight π_d for the dth component. Instead of maximizing the posterior probability given in Eq. (3.25), an approximation to it $q(\nabla x_p, k)$ is computed using variational method. The noise variance σ^2 is assumed to follow a Gamma distribution on the inverse of variance ($\Gamma(\sigma^{-2} \mid a, b)$ where a, b are the hyper-parameters). The cost function is the distance between the original distribution and the approximating one measured using KL-divergence between $p(\nabla x_p, k \mid \nabla y_p)$ and $q(\nabla x_p, k)$. The resulting cost is minimized to obtain k, by updating the parameters of the distributions alternately – parameters of one distribution is updated by marginalizing over the other distributions. The PSF is estimated as the mean of the marginal distribution of k. A multi-scale approach is used to estimate the PSF, wherein the estimation is performed by varying the PSF and image resolution from coarse to fine. The converged values at a coarser level is up-sampled and used as initialization for the next level. The full resolution kernel is obtained at the finest level. The estimated PSF is used to recover the image by using Richardson-Lucy algorithm. Though this method produces competitive results, it is computationally intensive.

Another single image motion deblurring algorithm is due to Jia [60] which utilizes the transparency created in the image due to motion blur. When an opaque object moves, the object boundaries partially occlude the background creating a semitransparent region around the object boundary. The transparency of each pixel is determined by the amount of time the background is exposed as the object moves. When the object is stationary the transparency values are either zero for the background pixels occluded by the body, or one. With object motion the transparency values deviate from being binary and take fractional values determined by the PSF and the unblurred object boundary shape. The PSF is estimated from the transparency values which in turn is estimated using MAP and refined using belief propagation. Richardson-Lucy algorithm is used to estimate the image once the PSF is obtained. User input is needed for collecting the foreground and background samples for transparency estimation.

Shan et al. [144] approach the problem of motion blur removal with a focus on removing the artifacts which are present in the results of both [36] and [60]. The authors use a better noise model and image prior for improving the results. They show that a simple Gaussian model for the noise does not capture the spatial randomness of the noise. Taking into consideration the error in the image and PSF estimate (Δx^i, and Δk^i, respectively), an intermediate image (I) is formed [144]

$$I = (x^i + \Delta x^i) \circledast (k^i + \Delta k^i) + n,$$

$$= x^i \circledast k^i + \Delta x^i \circledast k^i + x^i \circledast \Delta k^i + \Delta x^i \circledast \Delta k^i + n, \tag{3.26}$$

where x^i and k^i are the current estimates, n is the additive white Gaussian noise with standard deviation σ. From Eq. (3.26) it is seen that the error in PSF and image estimation becomes a part of the noise, if the noise model is not chosen appropriately. It is claimed in [144] that the ringing observed in the estimated image for most of the blind deconvolution algorithms is due to this mixing of noise and the error in estimation. In order to avoid this, a noise model which ensures spatial randomness is proposed. This uses the noise derivatives in the horizontal and vertical directions. Since noise is zero mean white Gaussian, the derivatives are also zero mean and they differ in the noise variance alone. For the nth order derivative the standard deviation of the noise is $\sqrt{2^n}\sigma$. Using the noise derivatives, the likelihood becomes

$$p(y|x,k) = \prod_{\partial^* \in \Theta} \prod_j N(\partial^* n_j | 0, \sigma_{o(\partial^*)}), \tag{3.27}$$

$$= \prod_{\partial^* \in \Theta} \prod_j N(\partial^* y_j | \partial^*(x \circledast k), \sigma_{o(\partial^*)}), \tag{3.28}$$

where ∂^* represents a partial derivative operator, $o(\partial^*)$, its order, and

$$\Theta = \{\partial^0, \partial_x, \partial_y, \partial_{xx}, \partial_{xy}, \partial_{yy}\},$$

with $\partial^0 n_j = n_j$, $\sigma_{o(\partial^*)}$ is the standard deviation of the derivative of order $o(\partial^*)$.

In addition to using the modified likelihood, [144] also uses a modified prior for the image, again with the purpose of reducing the ringing artifacts. The image prior is written as the product of a global prior and a local prior. Though the global prior used is same as that in [36], instead of using the prior on gradients directly, a piece-wise approximation is used for making the computation efficient. The local prior ensures that ringing is reduced by forcing similarity between the gradients of the estimated image and the observed image. Instead of doing it for the entire image, this is done in a locally smooth region so that the ringing in the estimated image is penalized. The kernel prior used was a mixture of Gaussians similar to [36].

The optimization is done using alternate minimization (AM). In each step of AM, methods like variable substitution and parameter re-weighting are used to improve the computation time and quality of result. The parameters which control the effect of the image prior are kept high initially leading to enhancement of strong edges and reduction of ringing. As iterations proceed these parameters are reduced making the likelihood more significant. This way the chances of converging to a trivial solution is eliminated. The results presented are better than that of [2, 36] and the computation time is also lesser.

An analysis of blind deconvolution methods with the view of explaining the apparent failure of MAP to produce non-trivial results was done by Levin et al. [2]. In [2] the conventional joint estimation of image and PSF by maximizing the posterior probability is indicated by $MAP_{x,k}$ and MAP_k is used to represent estimating PSF alone. It is argued that MAP_k solution is better than $MAP_{x,k}$ using the example of a scalar estimation. The priors used earlier are functions of image gradients and the authors in [2] provide a general form for this prior as

$$\log p(x) = -\sum_i (|g_{x,i}(x)|^\alpha + |g_{y,i}(x)|^\alpha) + C, \tag{3.29}$$

where $g_{x,i}(x)$ and $g_{y,i}(x)$ are the horizontal and vertical differences at pixel i, α determines the nature of the prior, with $\alpha < 1$ leading to sparse priors, $\alpha = 1$ a Laplacian prior and $\alpha = 2$ being Gaussian, C is a constant normalization term. A uniform prior is assumed for the PSF. $MAP_{x,k}$ solution is obtained by minimizing the cost

$$(\hat{x}, \hat{k}) = \arg\min_{x,k} \lambda \parallel x \circledast k - y \parallel^2 + \sum_i |g_{x,i}(x)|^\alpha + |g_{y,i}(x)|^\alpha. \tag{3.30}$$

Since the PSF prior is assumed to be uniform this term does not appear in the cost function. From Eq. (3.30), it is seen that for the trivial solution, which is $k = \delta(m, n)$, $x = y$, the data term goes to zero and the term corresponding to the prior forces the solution to a blurred one. This is described in more detail in Chap. 4. On the contrary, MAP_k solution estimates k by marginalizing over all possible images

$$\hat{k} = \arg\max_k p(k|y), \tag{3.31}$$

with

$$p(k|y) = \int p(x, k|y) dx. \tag{3.32}$$

The deconvolution process is demonstrated for the case of a Gaussian prior for the image. In [89] a general approach to solving the MAP_k for any prior using expectation maximization (EM) is provided by the same authors. Here the image prior used is a mixture of Gaussians. In the E-step, a non-blind deconvolution problem is solved to estimate the mean image given the current kernel and the covariance around it. In the M-step the kernel is estimated by taking into consideration the mean image and the covariance around the mean image. This distinguishes the MAP_k algorithm from the $MAP_{x,k}$ algorithm that considers only the mean image. A variational method is used to estimate the mean image and the covariance in the E-step, and the M-step involves solving a quadratic programming problem, the details of which are in [2]. Though the algorithm produces good deconvolution results, it is time consuming. Another observation that we make is that the analysis does not take into consideration the effect of PSF prior which, if considered, can avoid the problem of MAP giving trivial solutions. We address this issue in detail in Chap. 3.

Cho et al. [25] proposed a motion deblurring scheme which was designed for faster computation. This is an iterative method which estimates the image and the PSF alternately in each step of the iteration. The image and kernel estimation process was accelerated by introducing an intermediate prediction step and by working with image derivatives which are sparse. In the prediction step of the algorithm the gradient map of the estimated image is calculated. This predicts the salient edges in the image with noise suppression in smooth regions. At the beginning of the iteration the input to prediction step is the observed image and for further iterations the input is the image estimate of previous iteration. The prediction step consists of bilateral filtering to suppress noise and small details, followed by shock filtering which restores the strong edges. Since shock filter enhances noise, a gradient magnitude thresholding is performed to reduce the noise. These predicted gradient maps are used to estimate the kernel using FFTs which speed up the computation. Using this kernel the image is estimated with a Gaussian prior for the image. Though using a Gaussian prior leads to smoothed edges and ringing artifacts these are eliminated by the prediction step. Image estimation can also be done using FFTs, thereby speeding up the whole process. The image and kernel are updated in an iterative manner. The results are comparable to that of [2, 36, 144] with the added advantage that this method is many orders of magnitude faster than the previous algorithms.

3.5 Other Methods

In this section we discuss some of the blind deconvolution algorithms which do not fall into the above classes, but are notable for the quality of the estimated images. We also mention hardware based methods and multichannel blind deconvolution in this section, which are approaches quite different from those which were mentioned so far.

We discussed a few transform domain methods in the Sect. 3.1. Another transform domain method for blur identification is by using Radon transform [25, 116]. In [25], two methods for blind deconvolution are proposed, the first of which uses the edge information to construct the Radon transform of the blur kernel and uses inverse Radon transform followed by non-blind deconvolution. The second method proposed in [25] uses Radon transform in the MAP framework to estimate the image and blur simultaneously. These methods yield more accurate results when a lot of edges are available. Other motion blur estimation methods based on Radon transform have been reported in [72, 104, 116], with the work in [116] being an improvement over the other two methods.

Simulated annealing as a tool was used along with positivity and support constraints, in [97], for minimizing the energy of the error between the observed image and the estimated convolved image. The perturbation steps are repeated alternately for the image and PSF estimates, and the iteration proceeds till the global minimum is reached. The results demonstrated deconvolution of simple and synthetic images only. Lane [82] used the error measure in [97] with the image space constraints of positivity and support incorporated into the error measure. The resulting optimization problem was reformulated as an unconstrained minimization problem, which was solved using the conjugate gradient descent algorithm. The combined error function yields the stable convergence properties of Fienup's [127] error reduction algorithm. Here also the results were shown only for speckle images. Another algorithm on similar lines is the projection based blind deconvolution [169]. In [169], a priori knowledge of the image and PSF are expressed through constraint sets in the Hilbert space and a realizable projection based algorithm is proposed. But here also, the results are demonstrated only for speckle images.

Another class of blind deconvolution algorithms use a two step approach. The blur is estimated first, followed by a non-blind deconvolution. We describe a few algorithms which belong to this class. One of the earliest approaches to blur identification is reported in [13]. The blurs which could be identified are restricted to linear camera motion and out-of-focus lens system. The PSF in the linear camera motion case can be approximated by a rectangle whose orientation and length determine the direction and extent of the blur. In the case of out-of-focus lens system, the PSF is approximated by a disc whose radius determines the extent of blur for a circular aperture lens system. If the observation is noise free, the cepstrum of the observed image gives information regarding the above two blurs. In the case of motion blur, the periodic zeros (of periodicity $1/d$, where d is the extent of motion blur) of the Fourier transform of the PSF lead to a negative spike at a distance d from the origin. The direction of blur is given by the angle by which the spike has been rotated about the origin. For the out-of-focus PSF a ring of spikes is observed in the cepstrum. Once the blur parameters are estimated the PSF is estimated and deblurring is done. This algorithm is very sensitive to noise. Using bispectrum [18] for identifying frequency domain zeros gives better results at low signal to noise ratio since bispectrum has the ability to suppress additive, signal independent Gaussian noise. Though the bispectrum method works for parametric blurs with frequency domain zeros, other methods need to be devised for parametric blurs that do not have frequency domain zeros (eg. Gaussian blur).

Under the assumption of known noise variance and original image spectrum, Savakis et al. [139] proposed a blur identification method that estimates the blur using a collection of known blurs. The estimated PSF is that PSF from the collection which provides the best match between the restored residual spectrum and its original value.

Generalized cross validation (GCV) has been used for estimation of regularization parameter and stabilizing functional in [132]. Motivated by the success in [132] Reeves et al. used the method of GCV for blur identification in [131] and [133]. In [131] GCV was used to identify the parameters of the 2D-auto regressive moving average (ARMA) model, in which the image is modeled as an autoregressive (AR) process and the blur as a moving average (MA) process. This approach is perfected in [133] and compared with a maximum likelihood (ML) method for estimating the ARMA model parameters.

In [14], the PSF is estimated from one-dimensional Fourier analysis of the blurred image. The results presented are quite good, but the method works only for a limited class of blurs that can be expressed as a convolution of two-dimensional, symmetric, Levy-stable probability density functions. A Levy-stable probability density $(h(x, y))$ [14, 35] has a Fourier transform of the form

$$\hat{h}(\xi, \eta) = e^{-\alpha(\xi^2 + \eta^2)^\beta}, \quad \alpha > 0, 0 < \beta \le 1.$$

The estimated PSF is used to deblur the image, and both the steps use non-iterative methods. Another non-iterative method [62] gives a unique minimum norm solution provided an initial estimate for the image and the PSF are given. The algorithm works with some assumptions on the image and the PSF. The initial PSF estimate is assumed to be symmetric, positive and is normalized to unity. It makes additional assumptions on the observed image and the initial image estimate. The observed image is assumed to lie in the intersection of Wiener algebra (space of functions whose Fourier transform belongs to $\mathscr{L}(\mathbb{R})$) and $\mathscr{L}^2(\mathbb{R})$ space. The Fourier transform of the estimated image is assumed to be the product of Fourier transform of the observation and the sign of Fourier transform of the PSF. Under these restrictive assumptions, [62] provides a non-iterative method for blind deconvolution and also proves the existence and uniqueness of a minimum norm solution. i.e., a solution for the image and the PSF which has the smallest distance from the initial estimate of image and PSF, respectively. It is also proved that for the stated assumptions, if the observation satisfies a weak smoothness condition, the blind deconvolution becomes less ill-posed than the non-blind case. However the smoothness condition is satisfied only in the absence of noise. In the absence of noise the results show a considerable amount of deconvolution, giving images of good quality. The results are not satisfactory for the case of noisy observations. Periodic boundary conditions were assumed in [62]. The work in [62] was extended in [49] to cover the Neumann and anti-reflective boundary conditions.

Next we describe some hardware based solutions briefly. Cannon [57] uses an optical shift based image stabilizer (IS) which shifts the lens in parallel to the image plane to compensate for camera motion while imaging. Two shake detecting sensors are used which detect the angle and the speed of movement of the lens. This is used to generate drive signals for moving the IS lens to counteract the effect of shake. This method is effective only for relatively small exposures [111]. Liu et al. [168] have described an algorithm that makes use of the capability of CMOS sensors to capture multiple images within a normal exposure time. A motion detection algorithm is used at each pixel, and if any motion is detected temporal integration of the light field at the pixel is stopped. Another hardware based algorithm for motion blur removal [111] estimates the continuous PSF that caused the blur from sparse real time motion measurements taken during the exposure time. This PSF is then used to deblur the image by deconvolution.

There are other approaches also which include the multichannel blind deconvolution [43, 52, 125, 148–151], methods that use multiple images of the same scene [130, 178], and learning based methods [71, 170], the details of which are not discussed as the subject matter is not relevant to the topic of discussion of this monograph.

3.6 Discussions

From the previous sections, we notice that there exist several algorithms for the blind deconvolution problem. The recent ones focus on selecting better priors and using methods which estimate the PSF by marginalizing over all possible images. These algorithms give good performance only for a selected class of images, and the convergence issues are not explored. There still exist several open questions such as the suitability of joint estimation, existence of regularizers that can provide appropriate solutions and how accurate the solutions can be. Convergence analysis and the rate of convergence of iterative methods are other areas that need to be explored. In the following chapters we address the question of joint estimation, convergence analysis for the smoothness and TV regularizers and the design of a sparsity based regularizer.

Chapter 4
MAP Estimation: When Does It Work?

In this chapter we analyze the performance of the maximum a posteriori probability (MAP) method of estimating the image and the PSF, which we term as joint MAP estimation since both unknowns are estimated simultaneously. The MAP estimation procedure is explained in Sect. 2.4. Though the MAP and ML methods have been widely used in blind deconvolution, as can be seen from Chap. 3, recent work on camera shake removal [24, 36, 89, 144, 163, 179] shows a renewed interest in this area.

As we discussed in Chap. 3 many authors have reported the failure of direct application of the MAP in blind deconvolution, the details of which we give in the next section. Levin et al. [2] analyzed the reported failure of the MAP approach and showed that joint MAP estimation favors the blurred image as the solution when the image prior is a function of the image gradient. It is claimed in [2] that joint MAP gives a trivial solution of the estimated image as the blurred input and estimated PSF as the 2D discrete impulse, this being the global optimum of the MAP formulation. They suggest the use of MAP for estimating the blur alone by marginalizing over the image and then use a non-blind deconvolution to reconstruct the image.

The limitation of the work of Levin et al. [2] is that the authors assumed a uniform prior for the PSF. We prove that such a prior serves no purpose at all since it does not aid in preventing the trivial solution. On the contrary, we show that the MAP estimation does produce good results with an *appropriate* choice of PSF prior. The emphasis on the word appropriate is very important. We provide theoretical justification to show when joint MAP estimation works and when it does not. The arguments are substantiated through experimental results. We show how the regularization factor is to be selected so that the PSF regularization is effective.

This work is based on an earlier work: Joint MAP estimation for blind deconvolution: when does it work?, in Proceedings of the Eighth Indian Conference on Computer Vision, Graphics and Image Processing, ISBN 978-1-4503-1660-6, (2012-12-16) ©ACM, 2012. http://doi.acm.org/10.1145/2425333.2425383

© Springer International Publishing Switzerland 2014
S. Chaudhuri et al., *Blind Image Deconvolution: Methods and Convergence*,
DOI 10.1007/978-3-319-10485-0_4

Our analysis provides a feasible range for the regularization factor without using cross validation techniques [78, 91, 133]. We give an exact lower bound and an approximate upper bound for the PSF regularization factor. The work reported in [78, 91, 133] elucidates selection of regularization parameter using generalized cross validation (GCV). In addition to fixing the regularization parameter, Reeves et al. [133] use GCV as a method for blur identification. Liao et al. [91] use total variation (TV) as a prior for the image and the smoothness prior for the PSF and an alternate minimization (AM) is used for blind deconvolution. The regularization parameters here are estimated using GCV. Our analysis differs in usage of TV norm for both the image and the PSF, and also the range of acceptable regularization factor is obtained using an inexpensive computation, unlike the GCV.

Though we have already discussed the failure of MAP in Chap. 3, for maintaining continuity we briefly review it in the next section. The argument that Levin et al. [2] gave for explaining the failure of joint MAP is also discussed. Sect. 4.2 explains the limitation in the analysis of Levin et al. [2]. In addition, we prove that an appropriate choice of the PSF prior and PSF regularization factor prevents a trivial solution. We also derive the bounds for the PSF regularization factor. Experimental demonstrations and discussions follow in Sects. 4.4 and 4.5, respectively.

4.1 Reported Failure of Joint MAP

Recent work on camera shake removal from single image [36, 61, 85, 102, 144] have formulated the problem as a MAP estimation and make use of the fact that image gradient has a heavy tailed distribution. Since the image gradient distribution is used, all these solutions work in the image gradient domain and proceed by estimating the PSF and use it for deblurring the image using non-blind deconvolution algorithms. Fergus et al. [36] reported that direct application of the MAP method to estimate PSF and image fails. Their argument for the failure of MAP was that the MAP objective function attempts to minimize all gradients, whereas natural images are expected to have some strong gradients. This leads to a solution which is the trivial blurred image itself. Hence instead of estimating the image and PSF jointly using MAP (Sect. 3.3.2), an approximation to the posterior probability is computed followed by estimation of the PSF by maximizing the marginal probability. Essentially the blur kernel implied by a distribution of probable images is estimated as in [102].

Shan et al. [144] also uses a joint MAP for blind deconvolution, but with a better model to capture the spatial randomness of noise. This modifies the likelihood function in the gradient domain, and noise gradients up to second order are used to define the likelihood function. Another modification is that the image gradient prior is split as the product of a global and a local prior. The global prior for image gradients follows a logarithmic density [136, 165] and is approximated using two piece-wise continuous functions. The local prior is formed by constraining the blurred image gradient to be similar to the unblurred image gradient. The details of this method was already discussed in Chap. 3. In [144] and [36], additional components had to be used for the joint MAP formulation to give a non-trivial result. Levin et al. [2] address this problem and show that joint MAP gives only trivial solutions and is successful only for retrieving sharp edges. Their argument is summarized next.

The image formation model in Chap. 1 is repeated here for convenience.

$$y = k \circledast x + n, \tag{4.1}$$

where y denotes the noisy and blurred observation, k is the blur kernel which is assumed to be non-negative, x is the original image, $\circledast$ is the convolution operation, and n is additive white Gaussian noise.

Let $p(x, k|y)$ represent the posterior probability of x and k, once y is observed. Also let $p(x)$ and $p(k)$ represent the prior densities of x and k, respectively. In joint estimation using MAP, a pair $(\hat{x}, \hat{k})$ that maximizes $p(x, k|y)$ has to be obtained. Using the likelihood function and assuming that the image and blur are independent, the a posteriori probability can be written as

$$p(x, k|y) \propto p(y|x, k) p(x) p(k). \tag{4.2}$$

Most commonly used prior for the image is a function of the gradient of the image, and in general it can be expressed as [2]

$$\log p(x) = -\sum_i (|g_{h,i}(x)|^\alpha + |g_{v,i}(x)|^\alpha) + C, \tag{4.3}$$

where $g_{h,i}$ and $g_{v,i}$ denote the first-order horizontal and vertical difference, respectively, of the image at pixel location i, C is a normalizing constant, and α controls the sparsity of the image prior. Choice of $\alpha = 2$ gives the smoothness prior corresponding to a Gaussian Markov random field model. The optimum value for $(\hat{x}, \hat{k})$ is obtained by minimizing the negative of the logarithm of MAP, given by

$$(\hat{x}, \hat{k}) = \arg\min_{x,k} \left(\| k \circledast x - y \|^2 + \lambda_x \sum_i (|g_{h,i}(x)|^\alpha + |g_{v,i}(x)|^\alpha) - \lambda_k \log(p(k)) \right), \tag{4.4}$$

where λ_x and λ_k are the regularization parameters for the image and the PSF terms, respectively. Since MAP can be seen as regularized optimization, a more general form of cost function $C(\underline{x}, \underline{k})$ can be written as

$$C(\underline{x}, \underline{k}) = \| \underline{y} - K\underline{x} \|^2 + \lambda_x R_x(x) + \lambda_k R_k(k), \tag{4.5}$$

where $R_x(x)$ and $R_k(k)$ are the regularization factors for the image and the PSF, respectively, $\underline{x}$, $\underline{y}$, and $\underline{k}$ are the vectors obtained by lexicographic ordering, respectively, of x, y, and k. The suffix in the notation for regularizer indicates whether the regularizer corresponds to the image or the PSF and K denotes the convolution matrix corresponding to the PSF k.

In [2], the PSF is assumed to have a uniform distribution which makes the last term in Eq. (4.4) a constant, and we have

$$(\hat{x}, \hat{k}) = \arg\min_{x,k} \left(\parallel k \circledast x - y \parallel^2 + \lambda_x \sum_i (|g_{h,i}(x)|^\alpha + |g_{v,i}(x)|^\alpha) \right). \qquad (4.6)$$

To see the effect of a regularizer of the form Eq. (4.3) for the image and uniform prior for the PSF, we plot the cost of the R.H.S. of Eq. (4.6) for a blurred image and a sharp image in Fig. 4.1. The image is split into small blocks and the cost is evaluated in each of the blocks. In both the cases, the cost consists of only the term due to the regularizer. For the case where the image is sharp, i.e., the exact solution, the data term is zero, and for the trivial solution of k being the 2D-discrete impulse and $x = y$ also the data term is zero. Fig. 4.1 shows the cost for each block for the lena image for two different choices of block size: 50 and 10. From Fig. 4.1 it is seen that the cost introduced by the regularizer is smaller for the blurred image which is expected since blurring spreads out the edges leading to a reduction of the gradient values thereby reducing the cost. When the block size is small the effect of blurring is not as severe as for the case of large block size, which can be observed in Fig. 4.1b, where it can be seen that for a few blocks the cost for the blurred image is close to that of the sharp one. But in general, the cost of the blurred image is smaller compared to that of the sharp image. From this behaviour it is evident that using the gradient based regularizer for image and uniform prior for PSF leads only to trivial solution, since the minimization procedure will only select the blurred image as the solution which has lower cost.

It is argued in [2] that if $(\hat{x}, \hat{k})$ is a solution to Eq. (4.6), then a scaled version $(s\hat{x}, \hat{k}/s)$, with $s > 0$, also has the same value for the data term and the presence of the regularizer would choose the solution $\hat{x} \to 0$ and $\hat{k} \to \infty$, due to the second term corresponding to $p_x(x)$ in Eq. (4.6). It is suggested in [2] that k be chosen such that the mean value of the image is not changed, then

$$\sum_m \sum_n k(m, n) = 1, \text{with} k(m, n) \geq 0, \ \forall(m, n).$$

With this restriction on k, it is shown empirically in [2], that in a natural image, a trivial blurred solution is preferred by the joint MAP estimator, i.e., the estimated image and PSF are the observed image and the 2-D discrete impulse function, respectively. This happens since for the trivial solution, $(y, \delta(m, n))$, the data fitting term reduces to zero and the remaining image prior has lower cost for the blurred image than the sharp image. This observation is used to explain the failure of $MAP_{x,k}$, the joint solution, in [2]. To overcome the above limitation of $MAP_{x,k}$, Levin et al. [2, 89] propose a MAP based estimation of the PSF (MAP_k) first, followed by a non-blind deconvolution to estimate the image. Since MAP_k involves

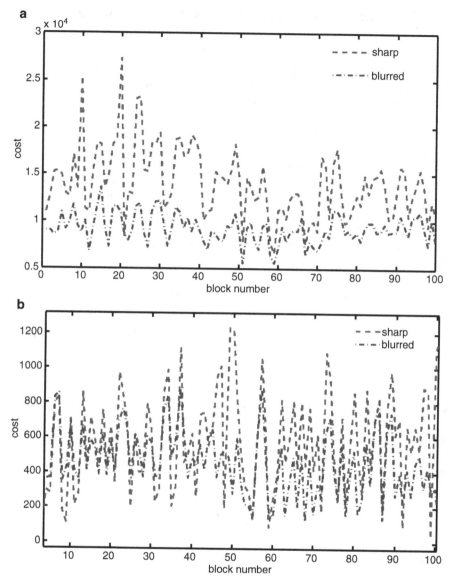

Fig. 4.1 Variation of cost equation (4.6) for the sharp and blurred images. (**a**) For block size of 50. (**b**) For block size of 10, with only 100 blocks shown

marginalization over all latent images, it is more challenging and is not explored much. In [89] a simple approximation to MAP_k is provided. This mainly differs from the existing alternate minimization techniques in the use of covariance of the latent image in addition to the mean image, which we have already discussed in Chap. 3.

4.2   Overcoming the Limitation of Joint MAP Estimation

We perform a joint estimation of the image and the PSF using a standard quadratic data term and commonly employed regularizers for the image and the PSF. The results show that the joint estimation does improve the quality of the image as can be seen visually as well as quantitatively through peak signal to noise ratio (PSNR) improvement. This observation is contradictory to the claim in [2].

We show that the claim in [2] is not valid, by observing that the shortcoming of the claim lies in choosing a uniform distribution for the PSF. Most PSFs are rarely uniformly distributed and they often show a structure such as radial or lateral symmetry. This is not true for the case of motion blur, yet in the case of motion blur the PSF is mostly sparse [179]. Even for motion blur, such an assumption (of uniform distribution) is inappropriate. We do use a non-uniform prior for the PSF which is represented by $R_k(\underline{k})$, as indicated in Eq. (4.5). We show that the use of a general prior, similar to Eq. (4.3), for the PSF prevents the trivial solution. Accordingly Eq. (4.4) becomes

$$(\hat{x}, \hat{k}) = \arg\min_{x,k} \| k \circledast x - y \|^2 + \lambda_x \sum_i (|g_{h,i}(x)|^\alpha + |g_{v,i}(x)|^\alpha) +$$
$$\lambda_k \sum_i (|g_{h,i}(k)|^\alpha + |g_{v,i}(k)|^\alpha). \tag{4.7}$$

The condition for avoiding the trivial solution is provided by Claim 4.1.

Claim 4.1. *The function* $R_k(k) = \sum_i (|g_{h,i}(k)|^\alpha + |g_{v,i}(k)|^\alpha)$ *attains the maximum when* k *is the impulse function, with* k *constrained to satisfy* $\sum_m \sum_n k(m,n) = 1$.

Proof. Consider the function

$$f(u, v) = \sum_{i=1}^N (|u_i|^\alpha + |v_i|^\alpha), \tag{4.8}$$

where u and v are length N vectors. Gradient of f is

$$\nabla f = [\alpha u_1^{\alpha-1}\text{sgn}(u_1), \cdots, \alpha u_N^{\alpha-1}\text{sgn}(u_N), \alpha v_1^{\alpha-1}\text{sgn}(v_1), \cdots, \alpha v_N^{\alpha-1}\text{sgn}(v_N)], \tag{4.9}$$

where $\text{sgn}(u_i)$ is the signum function defined as

$$\text{sgn}(u_i) = \begin{cases} 1 & u_i > 0, \\ -1 & u_i < 0. \end{cases}$$

From Eq. (4.9) it is seen that the gradient is discontinuous at $(\underline{0}, \underline{0})$. There is no stationary point for the function when $\alpha \leq 1$. In this case, the gradient does not exist at $(\underline{0}, \underline{0})$ neither does it go to zero at any other point. For $\alpha > 1$, there is one stationary point given by $(\underline{0}, \underline{0})$. It may also be noted that the function f in Eq. (4.8) is a sum of monotonically increasing functions and hence f is also a monotonically increasing function of (u, v). The minimum value of f is zero, which is attained at $(\underline{0}, \underline{0})$.

If u_i and v_i lie in the range $[0, 1]$, the maximum of the function occurs when $u_i = v_i = 1$. Comparing Eq. (4.8) with expression for $R_k(k)$, the maximum, $R_{k_{max}}$, of $R_k(k)$ occurs when the first-order difference in x and y directions becomes unity. Because of the constraints

$$\sum_m \sum_n k(m, n) = 1 \text{ and } k(m, n) \geq 0,$$

in the discrete case the first-order differences $|g_{h,i}(k)|$ and $|g_{v,i}(k)|$ both lie in the range $[0, 1]$, and become unity only when k is the 2-D discrete impulse function. $\square$

For illustration, we consider the total variation as the PSF regularizer, i.e., $R_k(k) = TV(k)$, where $TV(k)$ is defined as

$$TV(k) = \sum_i \sqrt{g_{h,i}^2(k) + g_{v,i}^2(k)}. \tag{4.10}$$

Using arguments similar to that in Claim 4.1 it can be proved that the behavior of Eq. (4.10) is similar to that of Eq. (4.8). We prove it below considering a general function $f(u, v)$.

$$f(u, v) = \sum_{i=1}^{N} \sqrt{u_i^2 + v_i^2}, \tag{4.11}$$

where u and v are length N vectors. Gradient of f is

$$\nabla f = \left[\frac{u_1}{\sqrt{u_1^2 + v_1^2}}, \cdots, \frac{u_N}{\sqrt{u_N^2 + v_N^2}}, \frac{v_1}{\sqrt{u_1^2 + v_1^2}}, \cdots, \frac{v_N}{\sqrt{u_N^2 + v_N^2}} \right]. \tag{4.12}$$

It may be noted from Eq. (4.12) that there is no stationary point for the function and that the function increases monotonically with a minimum value of zero at $(\underline{0}, \underline{0})$. The maximum value of the function $TV(k)$ (using forward difference) in this case is $R_{k_{max}} = 1 + \sqrt{1 + 1} + 1 = 3.414$.

From the above discussion, it is inferred that the necessary condition for preventing the trivial solution is a regularization term (in other words prior), which reaches its maximum when the PSF becomes the 2-D discrete impulse. But this is not sufficient to prevent a trivial solution. For the trivial solution (the no-blur solution), the estimated image is same as the observed image and the estimated PSF

is the discrete impulse. In this case the first term of Eq. (4.7) reduces to zero. The
second term has a lower value for the trivial solution as compared to the original,
since the trivial solution has weaker edges compared to the actual solution. Hence, to
avoid the trivial solution, it should be ensured that the term corresponding to the PSF
regularizer should be larger than the term corresponding to image regularizer, when
the PSF is the impulse function. This observation suggests that a proper choice of the
regularization factor λ_k in Eq. (4.7), prevents the trivial solution, thus eliminating
the bias of $MAP_{x,k}$ towards the blurry solution.

Hence, it is clear that λ_k should be chosen sufficiently large enough so that the
term from PSF regularizer is larger than the term arising from image regularizer.
This would define a lower bound for λ_k (say $\lambda_{k_{min}}$). It is important that λ_k
is also upper bounded, since choosing too large a value of λ_k forces excessive
smoothness on PSF leading to an averaging filter as the estimated PSF. We denote
the upper bound by $\lambda_{k_{max}}$. For a non-trivial solution, λ_k should be chosen such that
$\lambda_{k_{min}} < \lambda_k < \lambda_{k_{max}}$ as illustrated in Fig. 4.2. Thus we need to obtain the value for
$\lambda_{k_{min}}$ and $\lambda_{k_{max}}$.

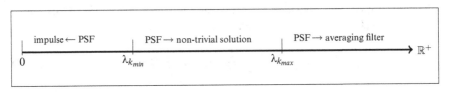

Fig. 4.2 Illustration of the effect of choice of λ_k on the estimated PSF. If $\lambda_k \in [0, \lambda_{k_{min}}]$,
estimated PSF tends trivially to an impulse function. If $\lambda_k \in (\lambda_{k_{min}}, \lambda_{k_{max}})$, a proper estimate
of both image and PSF is expected. If $\lambda_k > \lambda_{k_{max}}$, then PSF will again trivially approach the
averaging filter

Claim 4.2. *The lower bound of the PSF regularization factor, $\lambda_{k_{min}}$ is given by*

$$\lambda_{k_{min}} = \frac{\lambda_x E(R_x)}{R_{k_{max}}}.$$

*where $E(R_x)$ is the expectation of the image regularizer R_x and $R_{k_{max}}$ is the
maximum of $R_k(k)$.*

Proof. Since x is often modeled as a random field [58], $R(\underline{x})$ may also be treated
as a random variable. We fix a lower bound on λ_k by comparing the expectation of
image regularizer $(E(R_x))$ with $R_{k_{max}}$ for the trivial case. Using the condition that
the term due to PSF regularizer should dominate the term due to image regularizer
gives

$$\lambda_k > \frac{\lambda_x E(R_x)}{R_{k_{max}}}. \tag{4.13}$$

□

Since the minimum value of $R_k(k)$ is zero as seen from proof of Claim 4.1, it is not possible to upper bound using this minimum. From Fig. 4.2, we see that as λ_k increases beyond $\lambda_{k_{min}}$, the solution progressively tends towards producing an averaging kernel as the PSF. Using a rule of thumb to upper bound λ_k, the range of λ_k can be written as

$$\frac{\lambda_x E(R_x)}{R_{k_{max}}} < \lambda_k < \beta \frac{\lambda_x E(R_x)}{R_{k_{max}}}, \tag{4.14}$$

where $\beta > 1$ is an empirically chosen factor. As $\beta \to \infty$ the estimated PSF tends to the averaging filter. In this case, there is a deblurring but the solution is not an acceptable one. We show results which favor this explanation in the Sect. 4.4.

Here we have given the range in which the PSF regularization factor should lie. Regularization factor is usually estimated using a cross validation method. Cross validation based estimate of regularization factor requires optimizing a cost function with the regularization factor as the variable, making the computation very demanding. Our approach directly determines the range, based on the fact that the effect of PSF regularization must dominate the effect of image regularization when the PSF tends to a 2D discrete impulse, thus avoiding the computational overhead of cross validation techniques.

4.3 Choice of Image Priors

The previous section described the choice of PSF prior (regularizer) and the bounds on the PSF regularization factor. In this section, the image priors used in the experimentation are described. Two priors were considered for the image (i) the quadratic upper bounded total variation and (ii) the wavelet prior.

4.3.1 Quadratic Upper Bounded TV

We defined the quadratic upper bounded TV (Q_{TV}) in Sect. 2.3.2.2. For convenience we repeat the expression here

$$Q_{TV}(\underline{x}, \underline{x}^{(i)}) = \underline{x}^T D^T \Lambda^{(i)} D\underline{x} + K(\underline{x}^{(i)}). \tag{4.15}$$

$Q_{TV}(\underline{x}, \underline{x}^{(i)})$ is used as the image and PSF regularizer and since it depends on the previous value $\underline{x}^{(i)}/\underline{k}^{(i)}$, the cost function changes at each iteration. This necessitates using an outer loop to refine the estimates of $\underline{x}/\underline{k}$. Since alternate minimization is used the outer loop is by default taken care of by the AM algorithm. The data fitting term is the l_2-norm of the difference $\underline{y} - K\underline{x}$. With this data fitting term and approximated TV as the regularizers for both the image and the PSF, the cost function becomes

$$C_{TV}(\underline{x},\underline{k}) = \parallel \underline{y} - K\underline{x} \parallel^2 + \lambda_x \underline{x}^T D^T \Lambda_x^{(i)} D\underline{x} + \lambda_k \underline{k}^T D^T \Lambda_k^{(i)} D\underline{k}. \qquad (4.16)$$

The term $K(\underline{x}^{(i)})$ in the approximation for TV in Eq. (4.15) does not appear in Eq. (4.16) since it is a constant dependent on the previous iteration value and does not affect the optimization.

4.3.2 Wavelet Prior

As discussed in Sect. 2.3.1, a regularizer reduces the solution space by imposing some restrictions on the acceptable solution. Since wavelet transform has a good energy compaction property [94,112,162], natural images can be represented using a few large wavelet coefficients. This essentially means that we can consider the wavelet transform as a sparse representation of the image. Hence ℓ_1 norm of the wavelet coefficients is a suitable regularizer for image. We use the formulation by Vonesch et al. [162] for non-blind restoration for implementing the image estimation step of our algorithm. The idea used for non-blind restoration in [162] is briefly described here. The cost function for estimating the image is

$$C(\underline{x}) = \parallel \underline{y} - K\underline{x} \parallel^2 + \lambda \parallel Wx \parallel_1, \qquad (4.17)$$

where λ is the regularization factor and $\parallel Wx \parallel_1$ is the ℓ_1 norm of the wavelet coefficients of the image x. Minimizing Eq. (4.17) directly is difficult because of the convolution matrix. Instead, the cost function is replaced by a sequence of functions that are easier to minimize, such that the iterations lead to the minima of the cost function [29,162]. This is essentially the bound optimization principle [29] which is briefly described below.

Due to the presence of K, optimization of $C(\underline{x})$ is not easy. In bound optimization method, $C(\underline{x})$ is replaced by a succession of auxiliary functions which are easy to minimize [162]. Using the current estimate $\underline{x}^i$ an auxiliary function $C_i(\underline{x})$ is constructed which coincides with $C(x)$ when $x = x^i$ and upper-bounds $C(x)$ when $\underline{x} \neq \underline{x}^i$. $\underline{x}^i$ is updated as

$$\underline{x}^{i+1} = \arg\min_{\underline{x}} C_i(\underline{x}). \qquad (4.18)$$

A bound for Eq. (4.17) is obtained as [162]

$$C_i(\underline{x}) = \alpha \parallel \underline{x}^i - \underline{x} \parallel_2^2 + C(\underline{x}) - \parallel K\underline{x}^i - K\underline{x} \parallel_2^2, \qquad (4.19)$$

where α is a real positive scalar chosen as $\alpha > \rho(K^T K)$, and $\rho(K^T K)$ is the spectral radius of $K^T K$. Since K is the convolution matrix, this becomes the largest squared modulus of the DFT coefficients of K. It is shown in [162] that with some algebraic manipulations Eq. (4.19) can be written as

$$C_i(\underline{x}) = \| \underline{z}^i - \underline{x} \|_2^2 + \lambda\tau \| W\underline{x} \|_1, \qquad (4.20)$$

where $\underline{z}^i = \underline{x}^i + \alpha^{-1} K^T (\underline{y} - K\underline{x}^i)$, and $\tau = \alpha^{-1}$. The term $\underline{z}^i$ is a constant as far as the optimization process is concerned, since it depends only on the previous iterated value of $\underline{x}$, namely $\underline{x}^i$. Since the wavelet basis is orthonormal the quadratic term can be written in the wavelet domain as

$$C_i(\underline{x}) = \| W\underline{z}^i - W\underline{x} \|_2^2 + \lambda\tau \| W\underline{x} \|_1 . \qquad (4.21)$$

Equation (4.21) is decoupled and the minimization can be done coefficient-wise. The minimizer of Eq. (4.21) is

$$\underline{x}^{i+1} = W^T \mathscr{T}_{\gamma/2} W\underline{z}^i, \qquad (4.22)$$

where $\gamma = \lambda\tau$, and

$$\mathscr{T}_{\gamma/2}(w) = \mathrm{sgn}(w)(|w| - \gamma/2)_+,$$

where $(a)_+$ indicates $\max(a, 0)$. It is proved in [29], that the iterations converge to the minimum of $C(\underline{x})$. We use this in the image estimation step of AM with ℓ_1 norm of the wavelet transform as the image prior. The quadratic upper-bounded TV is used as the PSF regularizer. This gives the overall cost function as

$$C_{TV}(\underline{x}, \underline{k}) = \| \underline{y} - K\underline{x} \|^2 + \lambda_x \| W\underline{x} \|_1 + \lambda_k \underline{k}^T D^T \Lambda_k^{(i)} D\underline{k}. \qquad (4.23)$$

Alternate minimization is used to estimate the image and the PSF. At each iteration of AM, conjugate gradient descent algorithm is used to estimate the unknowns.

4.4 Experimental Demonstrations

Blind deconvolution using alternate minimization was carried out for joint estimation of the image and the PSF using the cost function in Eq. (4.5). Instead of using the total variation term as discussed in Sect. 4.3, a quadratic upper bound for TV was used [40] as the regularization term for PSF for computational convenience. It may be noted that the derivation done in Sect. 4.2 remains valid with the upper bounding, except that the range of the PSF regularization factor changes. For the image, we experimented with both the quadratic upper bounded TV (cost function in Eq. (4.16)) and the wavelet based regularizer (cost function in Eq. (4.23)) [39, 112, 162].

The results of deblurring for a noise variance of 10 is given in Fig. 4.3. The mask used here is a truncated Gaussian kernel of window size 5 with a spread $\sigma = 3$. The PSNR improvement obtained in this case is 3 dB. We show results for both the TV prior (Fig. 4.3b) as well as the wavelet prior (Fig. 4.3c) to demonstrate that TV need not be the only regularizer. The mean square error in the estimated PSF is 1.79×10^{-5}. It is seen that the estimated PSF could converge to three different values depending on the value of the PSF regularization term (λ_k). If λ_k is small, the effect of regularization for PSF is almost nil and the estimated PSF converges to impulse function and the estimated image is same as the original (Fig. 4.3d), which is as pointed out by Levin et al. [2]. If λ_k is large, the PSF converges to an averaging mask with total variation close to zero, leading to a relatively high mean square error of the PSF estimate (1.6×10^{-3}). This is shown in Fig. 4.3e, f, for the TV prior and wavelet prior respectively. For an intermediate value of λ_k the PSF converges to a value close to the original, with mean square error of the order of 10^{-5}.

Claim 4.2 in Sect. 4.2 is used as a guideline to select the appropriate value of λ_k. Since x is not known and only one observation y is available, we estimate $\lambda_{k_{min}}$ from a low pass filtered version of the observation y, i.e., we assume $E(R_x) \approx E(R_y)$. The low pass filtering of y is done before computing $E(R_y)$ to reduce the effect of noise shown in Eq. (4.1). λ_k is chosen in the range $(\lambda_{k_{min}}, \beta\lambda_{k_{min}})$ where β is a factor that is determined empirically. For the results shown in Fig. 4.4, the estimated $\lambda_{k_{min}}$ using $E(R_y)$ is 1.48×10^5 when an averaging filter of size 5 was used to denoise y. For $\lambda_k = 1.0 \times 10^5$ and 0.5×10^5, which are less than $\lambda_{k_{min}}$, the estimated PSF is a 2D discrete impulse which substantiates our claim. For $\lambda_k = 2 \times 10^5$, PSNR for restored image was the highest and the mean square error in PSF estimation was the least. For values of $\lambda_k \in (1.48 \times 10^5, 2 \times 10^5)$ and $\lambda_k \in (2 \times 10^5, 3 \times 10^7)$, a non-trivial result is obtained, but with a higher mean squared error for the PSF and lower PSNR for the image compared to $\lambda_k = 2 \times 10^5$. Experimentally obtained $\lambda_{k_{max}}$ is 3×10^7 and as λ_k tends to this value, the PSF becomes closer to an averaging filter PSF. In this case the value of β is close to 100, leading to non-trivial result for $\lambda_k \in (\lambda_{k_{min}}, 100\lambda_{k_{min}})$.

We also show the results on a different image which was blurred by a Gaussian kernel of window size 3, $\sigma = 1$ and with a Gaussian noise of variance 20 added to the blurred image. It is seen from Fig. 4.4d that when the regularization parameter is low, the estimated image is same as the observed image, and the estimated PSF in this case is the impulse function. For a properly chosen regularization term, even in the presence of high amount of noise (in this case noise variance is 20), a 3 dB improvement in PSNR is obtained (Fig. 4.4b). For large values of regularization parameter, the mean square error in PSF estimate is 1.3×10^{-4} (Fig. 4.4e, f).

Fig. 4.3 Results of deblurring for a 5×5 mask, and noise variance 10, using quadratic upper bounded TV prior the PSF. (**a**) Blurred and noisy image. (**b**) Deblurring result with $\lambda_{k_{min}} < \lambda_k < \lambda_{k_{max}}$; TV prior for image; PSNR improved by 3 dB. (**c**) Deblurring result with $\lambda_{k_{min}} < \lambda_k < \lambda_{k_{max}}$; wavelet prior for image; PSNR improved by 3 dB. (**d**) Result of deblurring for $\lambda_k < \lambda_{k_{min}}$, the result is almost same as the observed image and estimated PSF is an impulse function. (**e**) Deblurring result with $\lambda_k > \lambda_{k_{max}}$; TV prior for image; MSE for PSF is 1.6×10^{-3}. (**f**) Deblurring result with $\lambda_k > \lambda_{k_{max}}$; wavelet prior for image; MSE for PSF is 1.6×10^{-3}

Fig. 4.4 Results of deblurring for a 3×3 mask, and noise variance 20 using quadratic upper bounded TV prior for the PSF. (**a**) Blurred and noisy image. (**b**) Deblurring result with $\lambda_{k_{min}} < \lambda_k < \lambda_{k_{max}}$; TV prior for image; PSNR improved by 2.6 dB. (**c**) Deblurring result with $\lambda_{k_{min}} < \lambda_k < \lambda_{k_{max}}$; wavelet prior for image; PSNR improved by 3 dB. (**d**) Result of deblurring for $\lambda_k < \lambda_{k_{min}}$, the result is almost same as the observed image and estimated PSF is an impulse function. (**e**) Deblurring result with $\lambda_k > \lambda_{k_{max}}$; TV prior for image; MSE for PSF is 1.3×10^{-4}. (**f**) Deblurring result with $\lambda_k > \lambda_{k_{max}}$; wavelet prior for image; MSE for PSF is 1.3×10^{-4}

4.5 Discussions

We demonstrate that using an appropriate prior for the point spread function (PSF) a joint estimate for the image and the PSF can be reliably obtained using MAP. In addition we also provide a range of values for the PSF regularization parameter, for which a non-trivial solution is achieved. Estimating this range is computationally inexpensive. It is seen from the experiments that if the regularization factor is not high enough, indicating a weak regularization for PSF, the joint estimation fails, giving the observed image itself as the estimated image and impulse as the PSF. For large values of PSF regularization factor, the estimated PSF is the averaging filter which does give a non-negligible amount of deblurring, but with lower PSNR compared with that of a proper regularization factor. When the PSF regularization factor is chosen appropriately, a non-trivial solution with an improved PSNR is obtained, which validates our claim that joint MAP does work with appropriate regularization for PSF and a properly selected regularization factor. Having shown that joint MAP does give non-trivial solutions, we proceed with the convergence analysis of AM in the next two chapters.

Chapter 5
Convergence Analysis in Fourier Domain

In Chap. 4 we showed that joint estimation of the image and the PSF gives non-trivial results provided an appropriate PSF regularizer and the corresponding regularization factor is chosen. We observed in Chap. 2 that the blind deconvolution problem being a bilinear ill-posed one (Sect. 2.6), alternate minimization (AM) is a natural choice for estimating the image and the PSF using MAP or any other regularization based approaches. As seen earlier, AM generates a sequence of image, PSF pairs $\{(\hat{\underline{x}}_i, \hat{\underline{k}}_i)\}$, with i being the iteration number. In this chapter we do a frequency domain analysis to study the convergence of $\{(\hat{\underline{x}}_i, \hat{\underline{k}}_i)\}$. The only known convergence analysis of AM for blind deconvolution, has been carried out by Chan et al. [17] for a quadratic smoothness regularizer, which is proven to favor a very smooth solution. Our approach to circumvent the smoothness introduced by the quadratic regularizer is to use the TV regularizer, but this makes the analysis difficult. In order to make the convergence analysis feasible, we use the quadratic upper bounding approximation [40] to the TV norm for both the image and the point spread function (PSF). With this approximation, the cost function is quadratic at each step of the AM, thus making the problem amenable for convergence analysis.

There exist only few papers in literature that address the convergence of iterative methods for blind deconvolution. The work by Chambolle et al. [15] analyzes the minimization process for the case where the PSF is known. Figueiredo et al. [38] have studied the sufficiency conditions for convergence for the case of deconvolving images corrupted with Poisson noise, with the PSF assumed to be known. For the blind deconvolution case, the work by Chan et al. [17] analyzes the convergence of an AM method for the quadratic smoothness regularizer. In [106] it is shown that the blind deconvolution algorithm converges when the TV prior is approximated by a Gaussian distribution.

The convergence analysis by Chan et al. is summarized in Sect. 5.2.1. In the convergence analysis given by Chan et al. [17] for the smoothness norm, the solution to which the AM iterations converge depends on the initialization. The estimated image has the same phase as the phase of initial value, which is always the noisy

© Springer International Publishing Switzerland 2014
S. Chaudhuri et al., *Blind Image Deconvolution: Methods and Convergence*,
DOI 10.1007/978-3-319-10485-0_5

and blurred observation. The magnitude spectra of the estimated image and the PSF depend only on the magnitude spectrum of the observation and the regularization parameters. We note that this would mean that an iterative procedure is not needed to arrive at the solution! This also indicates that the convergence points could be derived in a non-iterative way; we show this in Sect. 5.2.2. We also provide an insight into the process that leads to deblurring in the case of smoothness, which also explains the better performance of TV norm. We extend the analysis of Chan et al. [17] and show that there is a gradual convergence to a better solution by using TV norm, and this would require an iterative updating of both the image and the PSF regularization factors. The solution is better in the sense that the iterative procedure makes the image and the PSF regularization factors depend on the image and PSF, respectively. In addition to the quadratic approximation to make the solution a linear one, we had to use a further approximation to make the system linear shift invariant at each iteration. The error term which arises due to this approximation is analyzed in Sect. 5.3. The implementation using conjugate gradient descent method and the results obtained are given in Sect. 5.4 followed by discussions in Sect. 5.5.

5.1   Deconvolution Framework

The image formation model in Eq. (4.1) is used, which can be written in matrix-vector form as

$$y = K\underline{x} + \underline{n}, \tag{5.1}$$

where the vector form of the images ($\underline{x}$ and $\underline{y}$) is obtained by lexicographically (i.e. converting the matrix to a vector by column wise or row wise ordering; we have used row wise stacking) ordering the 2-D images, $\underline{n}$ is the white Gaussian noise vector and K is the convolution matrix obtained from the PSF k. The PSF matrix is assumed to be of much smaller size than the image. In most practical cases, the PSF is non-negative and exhibits a lateral or radial symmetry. These properties of the PSF will be used as constraints along with the constraint that the sum of the PSF elements is normalized to unity to prevent any shift in the mean of the image.

Using a quadratic data fitting term and TV regularizer for the image and the PSF, the cost function obtained is

$$C(\underline{x}, \underline{k}) = \| y - K\underline{x} \|^2 + \lambda_x TV(\underline{x}) + \lambda_k TV(\underline{k}), \tag{5.2}$$

where $TV(.)$ is the total variation function. Using the quadratic upper bounding approximation of Sect. 4.3, the cost function in Eq. (5.2) gets modified to

$$C^i(\underline{x}, \underline{k}) = \| y - K\underline{x} \|^2 + \underline{x}^T D^T \Lambda_x^{(i)} D\underline{x} + \underline{k}^T D^T \Lambda_k^{(i)} D\underline{k}, \tag{5.3}$$

where i is the iteration index. $\Lambda^{(i)}$ is defined (for image as well as PSF) as

$$\Lambda^{(i)} = \mathrm{diag}(W^{(i)}, W^{(i)}), \tag{5.4}$$

where $\mathrm{diag}(L)$ refers to a diagonal matrix with elements of vector L as the diagonal and $W^{(i)}$ is a vector whose jth element is

$$w_j^{(i)} = \lambda_x \left(2\sqrt{(\Delta_j^h x^{(i)})^2 + (\Delta_j^v x^{(i)})^2} \right)^{-1}. \tag{5.5}$$

The cost function in Eq. (5.3) is minimized subject to the constraints

$$k(m,n) = k(-m,-n), \quad k(m,n) \geq 0 \quad \forall\, m,n,$$
$$\sum_m \sum_n k(m,n) = 1, \tag{5.6}$$

which come from symmetry, positivity of the PSF and mean invariance, respectively.

It is proved in [40] that $TV(\underline{x})$ and the quadratic approximation for TV Eq. (2.41) both lead to the same solution. The function in Eq. (5.3) is nonlinear in both the variables $\underline{k}$ and $\underline{x}$. In order to solve for $\underline{k}$ and $\underline{x}$, we use the alternate minimization method [16, 171] described in Sect. 2.6.

While estimating $\underline{x}$ at the $(i+1)$th iteration by keeping $\underline{k}$ fixed, the cost function is essentially

$$C_{\underline{x}}(\underline{x}) \triangleq C(\underline{x}, \underline{k})|_{\underline{k}=\underline{k}^{(i)}}$$
$$= \| \underline{y} - K^{(i)}\underline{x} \|^2 + \underline{x}^T D^T \Lambda_x^{(i)} D\underline{x} + B_1(\underline{k}^{(i)}), \tag{5.7}$$

where $B_1(\underline{k}^{(i)}) = \underline{k}^{(i)^T} D^T \Lambda_k^{(i-1)} D\underline{k}^{(i)}$, is a constant w.r.t. $\underline{x}$, and hence does not affect the optimization. An estimate of $\underline{x}^{(i+1)}$ is obtained by minimizing Eq. (5.7). Similarly the cost function for $\underline{k}$ given $\underline{x}$, is

$$C_{\underline{k}}(\underline{k}) \triangleq C(\underline{x}, \underline{k})|_{\underline{x}=\underline{x}^{(i+1)}}$$
$$= \| \underline{y} - X^{(i+1)}\underline{k} \|^2 + \underline{k}^T D^T \Lambda_k^{(i)} D\underline{k} + B_2(\underline{x}^{(i+1)}), \tag{5.8}$$

where X is the convolution matrix corresponding to the image x at the $(i+1)$th iteration. We have used the commutative property of circular convolution to write $K\underline{x} = X\underline{k}$. As in Eq. (5.7), B_2 depends only on $\underline{x}$, which is a constant when minimization is done w.r.t. $\underline{k}$, and hence can be neglected during the optimization process. Minimizing Eq. (5.8), gives $\underline{k}^{(i+1)}$. The two minimization steps are repeated till convergence.

The cost functions that appear in the AM steps are quadratic, as seen from Eqs. (5.7) and (5.8). Given $\underline{k}$, we solve for $\underline{x}$ by equating the gradient of Eq. (5.7) $(\nabla C_{\underline{x}})$ w.r.t. $\underline{x}$ to zero. This gives

$$(K_i^T K_i + D^T \Lambda_{\underline{x}}^{(i)} D)\underline{x} = K_i^T \underline{y}, \qquad (5.9)$$

Similarly, by equating $\nabla C_{\underline{k}}$ to zero, we solve for $\underline{k}$,

$$(X_{i+1}^T X_{i+1} + D^T \Lambda_{\underline{k}}^{(i)} D)\underline{k} = X_{i+1}^T \underline{y} \qquad (5.10)$$

It may be noted that Eqs. (5.9) and (5.10) are of the form $Ax = b$. It can be easily verified that in both cases, the matrix that corresponds to A is symmetric and positive semi-definite. This enables us to use conjugate gradient descent method to solve for $\underline{k}$ and $\underline{x}$. The number of iterations in the AM step is fixed empirically. The matrices K and D are of large size ($MN \times MN$ and $2MN \times MN$ respectively, where $M \times N$ is the image size) which makes it inefficient to create the matrices while implementing. Hence, the product Ax is evaluated using convolution (as will be explained in Sect. 5.4) and not by multiplication as written in Eqs. (5.9) and (5.10).

Let $\underline{x}^i$ and $\underline{k}^i$ be the estimated image and PSF, respectively, at the ith iteration of AM. As the AM iterations proceed it needs to be seen whether the sequences $\{\underline{x}^i | i = 0, 1, \cdots\}$ and $\{\underline{k}^i | i = 0, 1, \cdots\}$ converge to some value $\underline{x}^*$ and $\underline{k}^*$ such that $\| \underline{y} - K^* \underline{x}^* \|^2$ along with the regularization term is minimum. In the next section, we perform the convergence analysis and show that with proper choice of regularization parameters, the AM iterations do converge to a non-trivial solution.

5.2   Convergence Analysis of AM

We start by showing that the convergence analysis, as given in Chan et al. [17], could be done in an easier manner without using an iterative approach. We then provide the convergence analysis for the quadratic upper bounded TV and provide an explanation why it performs better than the smoothness regularizer. We also give a signal processing perspective to the convergence process.

5.2.1   Use of Quadratic Smoothness Regularizer

The convergence analysis for the quadratic smoothness regularizer was done by Chan et al. [17]. The cost function, with a quadratic data term and smoothness regularizer for both image and PSF, is

$$f(\underline{x}, \underline{k}) = \| \underline{y} - K\underline{x} \|^2 + \lambda_x r(\underline{x}) + \lambda_k r(\underline{k}), \tag{5.11}$$

where $r(\underline{x})$ is defined as

$$r(\underline{x}) = \sum_{m=0}^{M-2} \sum_{n=0}^{N-2} |\nabla x(m,n)|^2, \tag{5.12}$$

where $|\nabla x(m,n)|^2 = (x(m+1,n) - x(m,n))^2 + (x(m,n+1) - x(m,n))^2$ is the square of ℓ_2 norm of the gradient at (m,n).

Alternate minimization is used to minimize Eq. (5.11), and the two steps in an AM iteration correspond to solving the equations

$$(\overline{X}^i X^i + \lambda_k \Delta)\underline{k}^{i+1} = \overline{X}^i \underline{y} \tag{5.13}$$

$$(\overline{K}^{(i+1)} K^{i+1} + \lambda_x \Delta)\underline{x}^{i+1} = \overline{K}^{(i+1)} \underline{y}, \tag{5.14}$$

where $\overline{X}$ is the conjugate transpose of X (for a general X), and Δ is the 2D discrete Laplacian operator which is block circulant with circulant blocks (BCCB). The convergence analysis in [17] proceeds by taking the Fourier transforms of Eqs. (5.13) and (5.14) which give an iterative equation for the image and the PSF in the transform domain. Denoting the Fourier transform of $\underline{x}$, $\underline{k}$, and $\underline{y}$ by $\mathscr{X}$, $\mathscr{K}$, and $\mathscr{Y}$ respectively, the iterative expression for the Fourier transforms are

$$\mathscr{K}^{i+1}(\xi_x, \xi_y) = \frac{\overline{\mathscr{X}^i(\xi_x, \xi_y)}\mathscr{Y}(\xi_x, \xi_y)}{|\mathscr{X}^i(\xi_x, \xi_y)|^2 + \lambda_k \mathscr{R}(\xi_x, \xi_y)}, \tag{5.15}$$

$$\mathscr{X}^{i+1}(\xi_x, \xi_y) = \frac{\overline{\mathscr{K}^{i+1}(\xi_x, \xi_y)}\mathscr{Y}(\xi_x, \xi_y)}{|\mathscr{K}^{i+1}(\xi_x, \xi_y)|^2 + \lambda_x \mathscr{R}(\xi_x, \xi_y)}, \tag{5.16}$$

where $\overline{\mathscr{X}^i(\xi_x, \xi_y)}$ is the complex conjugate of $\mathscr{X}^i(\xi_x, \xi_y)$ and $\mathscr{R}(\xi_x, \xi_y)$ is the Fourier domain representation of the 2D discrete Laplacian Δ, given by

$$\mathscr{R}(\xi_x, \xi_y) = 4 - 2\cos\frac{2\pi\xi_x}{M} - 2\cos\frac{2\pi\xi_y}{N}. \quad \xi_x = 0 \cdots M-1, \xi_y = 0 \cdots N-1, \tag{5.17}$$

which is plotted in Fig. 5.1 and corresponds to a high pass filter.

The plot shows a centered spectrum, with the center corresponding to low frequency and corners represents high frequencies. From Eqs. (5.15) and (5.16), the phase and magnitude convergences are analyzed separately. It is shown that the phase of the estimated image is same as the phase of the initial value of the image. The phase of the estimated PSF is zero provided the initial value of the image is same as the blurred and noisy observation. The convergence of magnitude is as given in Lemma 5.1 and we repeat the proof in [17] here for completeness.

Lemma 5.1. *The magnitude spectra of the PSF and the image converge to*

$$M_S = \sqrt{\sqrt{\frac{\lambda_k}{\lambda_x}}|\mathscr{Y}| - \lambda_k \mathscr{R}}$$

and $\sqrt{\dfrac{\lambda_x}{\lambda_k}} M_S$, *respectively, when* $|\mathscr{Y}|^2 > \lambda_x \lambda_k \mathscr{R}^2$, *else they both converge to zero.*

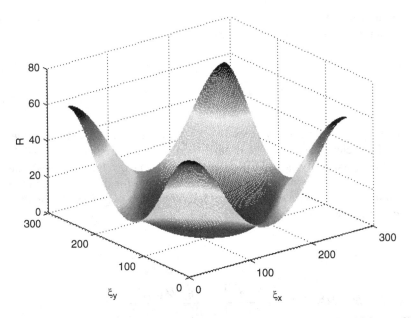

Fig. 5.1 Spectral representation of the 2D discrete Laplacian kernel, which is a high pass filter. The spectrum is centered, with the center corresponding to low frequency and corners representing high frequencies

Proof. We use u, v, and w to represent the magnitudes of $\mathscr{X}$, $\mathscr{K}$, and $\mathscr{Y}$, respectively. Also, let $r = \mathscr{R}(\xi_x, \xi_y)$. Using these in Eqs. (5.15) and (5.16) gives

$$v_{i+1} = \frac{u_i w}{u_i^2 + \lambda_k r}, \tag{5.18}$$

$$u_{i+1} = \frac{v_{i+1} w}{v_{i+1}^2 + \lambda_x r}. \tag{5.19}$$

Substituting Eq. (5.19) in Eq. (5.18), we get the update equation for v as

$$v_{i+1} = \frac{v_i w^2 (v_i^2 + \lambda_x r)}{v_i^2 w^2 + \lambda_k r (v_i^2 + \lambda_x r)} \tag{5.20}$$

From Eq. (5.20), the update equation for v becomes $v_{i+1} = F(v_i)$, where $F(.)$ is defined as

$$F(v) = \frac{v w^2 (v^2 + \lambda_x r)}{v^2 w^2 + \lambda_k r (v^2 + \lambda_x r)} \tag{5.21}$$

A similar update equation can be written for u. Since the expressions are similar, the analysis is done for only one of the variables say v. The update equation leads to a convergent series if $F(.)$ is a contractive mapping and the point of convergence corresponds to the fixed point of $F(.)$. Solving $F(v) = v$, gives the fixed points of the function as

$$v_1 = 0, \quad \text{and} \quad v_2 = \sqrt{\sqrt{\frac{\lambda_x}{\lambda_k}} w - \lambda_x r} \tag{5.22}$$

The convergence properties depend on the range in which w value lies. It can take values in three ranges which are determined by the stationary points and fixed points of the function.

Case 1: $w < \sqrt{\lambda_x \lambda_k} r$
If $w < \sqrt{\lambda_x \lambda_k} r$, there is only one fixed point, namely $v_1 = 0$. The plot of $F(v)$ for this case is as shown in Fig. 5.2. Since the fixed point is zero in this case,

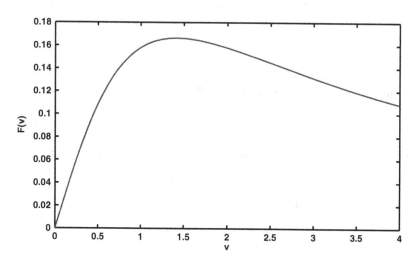

Fig. 5.2 Plot of $F(v)$ when $w < \sqrt{\lambda_x \lambda_k} r$

the iterations converge to zero provided the function is contractive in nature. The function $F(v)$ is contractive if $F(v) < v$, $\forall v$, which leads to the condition for being contractive as $|F'(v)| < 1$. From Eq. (5.21), $F'(v)$ is

$$F'(v) = \frac{w^2(v^2 - \lambda_x r)}{v^2 w^2 + \lambda_k r(v^2 + \lambda_x r)^2} \frac{(v^2 w^2 - (v^2 + \lambda_x r)^2 \lambda_k r)}{v^2 w^2 + \lambda_k r(v^2 + \lambda_x r)^2} \tag{5.23}$$

Using simple algebraic manipulations on Eq. (5.23) it can be shown that

$$|F'(v)| < \frac{w^2}{\lambda_x \lambda_k r^2} < 1, \qquad \text{since } w < \sqrt{\lambda_x \lambda_k} r. \tag{5.24}$$

From Eq. (5.24), $F(v)$ is contractive and the iterations converge to the fixed point zero when $w < \sqrt{\lambda_x \lambda_k} r$. For $w > \sqrt{\lambda_x \lambda_k} r$, there are two fixed points v_1 and v_2. To analyze the convergence properties when $w > \sqrt{\lambda_x \lambda_k} r$, we split the range into two, based on the stationary points of $F(v)$ which are given by

$$v_a = \sqrt{\lambda_x} r,$$

$$v_b = \frac{w}{\sqrt{2\lambda_k r}} + \frac{\sqrt{w^2 - 4\lambda_x \lambda_k r^2}}{\sqrt{2\lambda_k r}},$$

$$v_c = \frac{w}{\sqrt{2\lambda_k r}} - \frac{\sqrt{w^2 - 4\lambda_x \lambda_k r^2}}{\sqrt{2\lambda_k r}}. \tag{5.25}$$

Case 2: $\sqrt{\lambda_x \lambda_k} r < w < 2\sqrt{\lambda_x \lambda_k} r$

If $w^2 < 4\lambda_x \lambda_k r^2$, there exists only one stationary point namely v_a. For $\sqrt{\lambda_x \lambda_k} r < w < 2\sqrt{\lambda_x \lambda_k} r$, the function has two fixed points v_1 and v_2. It may be noted that the fixed point v_2 (in Eq. (5.22)) is less than the stationary point v_a (in Eq. (5.25)). The fixed point v_2 and the stationary point v_a define three ranges for w as shown in Fig. 5.3.

Region I $[v_1, v_2]$. Here the function is increasing and $F(v) > v$ and the iterations converge to fixed point v_2.

Region II $[v_2, v_a]$. Since the slope of the function in this region is positive, in this region also the function increases but with $F(v) < v$, i.e., the function is a contractive mapping and since the fixed point v_2 is an element of this range, the iterations converge to v_2.

Region III $[v_a, \infty)$. In this region the function is decreasing, and at a $v_0 \in [v_a, \infty)$, the function has value $F(v_0) < F(v_a)$, which means that the range $[v_a, \infty)$ gets mapped to the range $[0, v_a]$. Since this corresponds to Region I and Region II, the iterations converge to the fixed point v_2 as explained above.

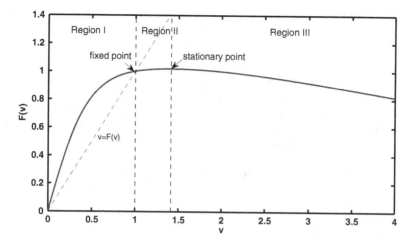

Fig. 5.3 $F(v)$ when $\sqrt{\lambda_x \lambda_k} r < w < 2\sqrt{\lambda_x \lambda_k} r$

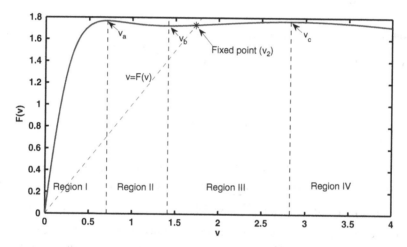

Fig. 5.4 $F(v)$ when $w > 2\sqrt{\lambda_x \lambda_k} r$

The iterations converge to the fixed point $v_1 = 0$ only when the initial value of v is chosen as zero, else it always converges to $v_2 = \sqrt{\sqrt{\dfrac{\lambda_x}{\lambda_k}} w - \lambda_x r}$, when w is in the range $\sqrt{\lambda_x \lambda_k} r < w < 2\sqrt{\lambda_x \lambda_k} r$.

Case 3: $w > 2\sqrt{\lambda_x \lambda_k} r$.

In this case the function has two fixed points v_1 and v_2, and three stationary points v_a, v_b, and, v_c. The plot of the function is shown in Fig. 5.4. The stationary points split w into four regions and the convergence behaviour is as follows:

Region I $[0, v_c]$. The function is increasing and $F(v) > v$ in this region. Hence
any element $v_0 \in [0, v_c]$ gets mapped either to Region I, II, or III.

Region II $[v_c, v_a]$. Here also $F(v) > v$, and for large enough values of w,
elements of this range gets mapped into Region III.

Region III $[v_a, v_b]$. The fixed point v_1 is contained in this region. The function
is increasing with $F(v) > v$, for $v \in [v_a, v_2]$ and $F(v) < v$, for $v \in [v_2, v_b]$.
In both the cases, the iterations converge to v_1. We saw that points in Region
I and II gets mapped to Region III, hence a point which lies in any of the
regions I to III, eventually converge to the fixed point v_2.

Region IV $[v_b, \infty)$. Here the function is decreasing and also $F(v) < v$ and a
point in this region gets mapped back to regions I, II, or III. Since a point in
this region eventually converge to the fixed point v_1, it follows that a point in
Region IV also converges to v_1.

From the above set of arguments it is seen that unless otherwise the initial value
of v is selected as zero, the iterations always converges to the fixed point v_1 when
$w > \sqrt{\lambda_x \lambda_k} r$.

Substituting the value at convergence of v, $v = \sqrt{\sqrt{\dfrac{\lambda_x}{\lambda_k}} w - \lambda_x r}$ in Eq. (5.19),

the value of u at convergence is obtained as

$$u = \sqrt{\frac{\lambda_k}{\lambda_x}} \sqrt{\sqrt{\frac{\lambda_x}{\lambda_k}} w - \lambda_x r} \qquad (5.26)$$

□

As pointed out in [17], one of the disadvantages of smoothness regularizer
(as seen from Lemma 5.1) is that the magnitude spectrum of the image and the PSF
differ only by a constant factor $\sqrt{\dfrac{\lambda_x}{\lambda_k}}$, which is almost never true in practice. Hence
the solution is a trivial one and does not achieve deconvolution in general. In our
analysis of the quadratic upper bounded TV we show that this does not happen
with the TV norm, and that we do obtain a non-trivial solution, which is one reason
why TV performs better than smoothness regularizer.

From Lemma 5.1, it is seen that the convergence is point wise at each frequency.
If the inequality in the Lemma 5.1 is not satisfied, at any frequency, the estimated
image magnitude at that point goes to zero. The implications of this is discussed in
the next section.

The convergence analysis has an elegant explanation from a signal processing
point of view, which we elaborate next. This helps us to see how to select the
regularization terms λ_x and λ_k. Using this we also argue why TV norm performs
better. We also show that the convergence points derived in Chan et al. [17] can also
be derived without using an iterative procedure.

5.2.2 Illustration of Point-Wise Convergence

Rewriting the cost function in Eq. (5.11) in frequency domain with smoothness as the regularizer, and making use of Parseval's theorem, leads to

$$C_f(\mathscr{X}, \mathscr{K}) = \| \mathscr{Y} - \mathscr{K}.\mathscr{X} \|^2 + \lambda_x \| \mathscr{R}.\mathscr{X} \|^2 + \lambda_k \| \mathscr{R}.\mathscr{K} \|^2, \qquad (5.27)$$

where $\mathscr{R}.\mathscr{X}$ is the Fourier transform of the smoothness regularizer, '.' indicates point wise multiplication, and $\mathscr{R}.\mathscr{K}$ is defined similarly. We have dropped the frequency indices (ξ_x, ξ_y) for simplicity. The smoothness regularizer defined in Eq. (5.12) can be written in a matrix-vector product form as

$$r(\underline{x}) = \sum_{m=0}^{M-2} \sum_{n=0}^{N-2} |\nabla x(m,n)|^2, \qquad (5.28)$$

$$= \underline{x}^T D^T D \underline{x} = \| D \underline{x} \|_2^2. \qquad (5.29)$$

where D is the derivative operator defined as $D = [(D^h)^T \ (D^v)^T]^T$. D^h and D^v denote matrices such that $D^h \underline{x}$ and $D^v \underline{x}$ are the vectors of all horizontal and vertical first-order differences of the vector x, respectively. From Eq. (5.28) it is seen that the Fourier transform can be written as $\mathscr{R}.\mathscr{X}$. As convolution becomes multiplication in Fourier domain, the variables are separated and the cost function could be seen as a sum of terms of the form

$$f(s_1, s_2) = (t - s_1 s_2)^2 + \lambda_x s_1^2 + \lambda_k s_2^2, \qquad (5.30)$$

with t, s_1 and s_2 assumed to be real, non-negative scalar quantities. Since the variables are separated, the minima for each $(\mathscr{X}, \mathscr{K})$ pair is obtained by minimizing each of the terms individually. We look at the minima of Eq. (5.30).

We show that the function $f(s_1, s_2)$, has minima similar to that of the problem addressed by Chan et al. [17]. It may be noted that this function is convex for non-negative values of s_1 and s_2. This function has a single, real, positive, non-trivial stationary point, given by

$$s_1 = \sqrt{\sqrt{\frac{\lambda_k}{\lambda_x}} t - \lambda_k}, \qquad (5.31)$$

$$s_2 = \sqrt{\frac{\lambda_x}{\lambda_k}} \sqrt{\sqrt{\frac{\lambda_k}{\lambda_x}} t - \lambda_k}. \qquad (5.32)$$

Note that the point $s_1 = 0$, $s_2 = 0$ is also a stationary point. The Hessian of Eq. (5.30) at the non-zero stationary point is given by

$$
\Sigma = 2 \begin{bmatrix} \sqrt{\dfrac{\lambda_x}{\lambda_k}}t & t - 2\sqrt{\lambda_x \lambda_k} \\[3mm] t - 2\sqrt{\lambda_x \lambda_k} & \sqrt{\dfrac{\lambda_k}{\lambda_x}}t \end{bmatrix}. \tag{5.33}
$$

The condition for the non-zero stationary point to be a minimum is that the Hessian at that point is positive definite [26]. We use Sylvester condition [26] for checking positive definiteness, according to which a real symmetric matrix is positive definite if and only if all the leading principal minors are positive. Since the first element of Σ is positive, the condition for the non-zero stationary point to be the minima is

$$
t^2 - (t - 2\sqrt{\lambda_x \lambda_k})^2 > 0,
$$

$$
t > \sqrt{\lambda_x \lambda_k}. \tag{5.34}
$$

If Eq. (5.34) is not satisfied, the only minimum is at the point $(0,0)$. When Eq. (5.34) is satisfied, since s_1 and s_2 are assumed to be non-negative, there exists only one minimum, which is the positive value in Eq. (5.31). Using this in Eq. (5.27), it could be seen that at each point in frequency, a minimum of the form Eq. (5.31) is obtained with t, $\lambda_x s_1$, $\lambda_k s_2$ replaced by $|\mathscr{Y}|$, $\lambda_x \mathscr{R}$, $\lambda_k \mathscr{R}$ respectively. So at each point, depending on the magnitude of the input image spectrum at that point, the estimated magnitude spectrum converges to either zero or to Eq. (5.31). The condition for convergence in this case is similar to that of Eq. (5.34), but modified as

$$
|\mathscr{Y}| > \sqrt{\lambda_x \lambda_k} \mathscr{R} \tag{5.35}
$$

to include the regularization term, as can be seen from Eq. (5.27). The effect of the value of the regularization parameters on the reconstructed spectrum is shown in Fig. 5.5. For high values of regularization parameters the high frequency components of the spectrum is reduced to zero and hence edge information is lost leading to a blurred output. If the values of regularization parameters are low, the high frequency components are also passed and there is hardly any denoising. For the smoothness regularizer, $\mathscr{R}(\xi_x, \xi_y)$ is the frequency dependent function shown in Fig. 5.1. Three key observations could be made from Fig. 5.5,

1. The R.H.S. of the inequality Eq. (5.35) is frequency dependent. Since $\mathscr{R}(\xi_x, \xi_y)$ is small for low frequencies as can be seen in Fig. 5.1, the comparison threshold is low at lower frequencies and it increases at high frequencies.
2. At each frequency, the reconstructed spectrum depends on the difference of the scaled observed magnitude spectrum and scaled $\mathscr{R}(\xi_x, \xi_y)$. So at low frequencies, provided the inequality Eq. (5.35) is satisfied, the magnitude converges to a value close to the observed magnitude. At higher frequencies, since $\mathscr{R}(\xi_x, \xi_y)$

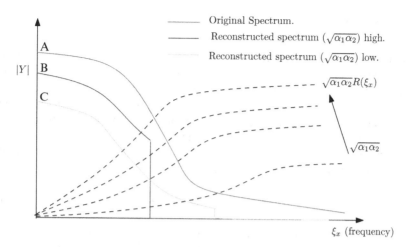

Fig. 5.5 Illustration of the effect of regularization parameter on the estimated image magnitude spectrum. The plot A corresponds to the spectrum of original image (x). The plot B is the estimated spectrum with a high regularization factor when high frequency components are lost. The plot C corresponds to a smaller amount of regularization when more high frequency components are retained at the cost of being quite away from the true spectrum. The *dotted curves* correspond to the high pass filter equivalent to the regularizer

is large, convergence is to a value smaller than the observed magnitude, which reduces the noise, but could also force high frequency components corresponding to edges to go to zero.

3. For a given frequency, depending upon the value of $\sqrt{\lambda_x \lambda_k}$, the performance varies. For high values of $\sqrt{\lambda_x \lambda_k}$ the high frequency components are forced to zero, leading to loss of edge information, and ringing in the image due to Gibb's phenomenon. Whereas if $\sqrt{\lambda_x \lambda_k}$ is low, then high frequency components are passed, leading to a noisy result without any deblurring, which means that the estimated PSF tends to a delta function. It may be noted that Levin et al. [88] also point to a similar effect while doing a maximum a posteriori probability (MAP) based joint estimation of image and PSF, for the case when the regularization term for PSF is not strong enough. This matches with our observation in Chap. 3 that the regularization factor should be chosen to be greater than $\lambda_{k_{min}}$.

We will next show in our analysis of the quadratic upper bounded TV that the estimated image and PSF magnitudes, though related by a scale factor, are better compared to the solution in [17], since the scaling factors are signal (image or PSF) dependent. By approximating the quadratic upper bounded case further, we get a situation similar to that of Chan et al. [17], with the key difference that λ_x and λ_k vary with each iteration, leading to an iterative convergence.

5.2.3 Use of TV Regularizer

We now present a Fourier domain convergence analysis of the alternate minimization procedure applied to Eq. (5.3). For analysis purpose we have assumed circular convolution which is commutative in nature i.e., $K\underline{x} = X\underline{k}$, where X is the convolution matrix corresponding to $\underline{x}$. The vector $\underline{k}$ is padded with zeros to make its size same as that of $\underline{x}$. It may be noted from Eqs. (5.7) and (5.8) that the cost functions for estimating $\underline{x}$ for a given $\underline{k}$ and $\underline{k}$ given $\underline{x}$ are quadratic in nature. The nature of the joint cost function Eq. (5.3) is, however, not quadratic due to the first term, which is reproduced here for convenience

$$\| \underline{y} - K\underline{x} \|^2 = \underline{y}^T \underline{y} - 2\underline{y}^T K\underline{x} + \underline{x}^T K^T K\underline{x}. \tag{5.36}$$

Following the method of [17] the convergence analysis is done in the frequency domain, by taking the discrete Fourier transform (DFT) of Eqs. (5.9) and (5.10). K and X being convolution matrices, are block circulant with circulant blocks (BCCB) and are diagonalized by the DFT as indicated in Eqs. (5.37) and (5.38). Let F_N denote a Fourier transform matrix of size $N \times N$. The BCCB matrices K and X are of size $MN \times MN$. Then

$$K = (F_M \otimes F_N)\Lambda_k(F_M \otimes F_N)^*, \tag{5.37}$$

$$X = (F_M \otimes F_N)\Lambda_X(F_M \otimes F_N)^*, \tag{5.38}$$

where $\otimes$ indicates the Kronecker product and Λ_k is a diagonal matrix whose diagonal entries correspond to the Fourier transform of $\underline{k}$, i.e.,

$$\mathrm{diag}(\Lambda_k) = (F_M \otimes F_N)\underline{k}. \tag{5.39}$$

Similarly

$$\mathrm{diag}(\Lambda_X) = (F_M \otimes F_N)\underline{x}. \tag{5.40}$$

Let $\mathscr{X}(\xi_x, \xi_y)$ and $\mathscr{K}(\xi_x, \xi_y)$ represent the DFT of X and K, respectively. Equations (5.39) and (5.40) give $\mathscr{K}$ and $\mathscr{X}$, respectively.
Since $D = [D^{h^T} \ D^{v^T}]^T$

$$D^T \Lambda_k^{(i)} D\underline{k} = D^{h^T} \Lambda_{kh}^{(i)} D^h\underline{k} + D^{v^T} \Lambda_{kv}^{(i)} D^v\underline{k}, \tag{5.41}$$

$$D^T \Lambda_x^{(i)} D\underline{x} = D^{h^T} \Lambda_{xh}^{(i)} D^h\underline{x} + D^{v^T} \Lambda_{xv}^{(i)} D^v\underline{x}, \tag{5.42}$$

where $\Lambda_{kh} = \Lambda_{kv} = \mathrm{diag}(W^{(i)})$, with W as defined in Eq. (5.5) and evaluated for the PSF. Λ_{xh} and Λ_{xv} are defined similarly using the W for image. The terms $D^{h^T} \Lambda_{kh}^{(i)} D^h$ and $D^{v^T} \Lambda_{kv}^{(i)} D^v$ in Eq. (5.41) are unfortunately not BCCB, but the matrices D^v and D^h which correspond to convolution matrix of the difference operation are BCCB, and are diagonalized by $F_m \otimes F_n$. Since D^h is the convolution matrix corresponding to the first order horizontal difference, there exists an equivalent convolution mask given by $h_1 = [0, 1, -1]$. Similar convolution masks can be associated with D^{h^T}, D^v, and D^{v^T} denoted by h_2, h_3 and h_4, respectively, defined as

$$h_2 = [-1, 1, 0],$$

$$h_3 = [0, 1, -1]',$$

$$h_4 = [-1, 1, 0]'.$$

From a system perspective, the matrix $D^{h^T} \Lambda_{kh}^{(i)} D^h$ operating on the vector $\underline{k}$ is equivalent to two convolutions and a multiplication in spatial domain, given by

$$D^{h^T} \Lambda_{kh}^{(k)} D^h \underline{k} \equiv h_2 \circledast (W(m, n).(h_1 \circledast k)), \tag{5.43}$$

where $\circledast$ denotes the convolution operation, '.' indicates point-wise multiplication and $W(m, n)$, $m \in [0 \cdots M]$, $n \in [0 \cdots N]$, is the matrix form of W as defined in Eq. (5.5). This system is shown in Fig. 5.6 and it may be noted that due to the multiplication by W, the system though linear, is not shift invariant.

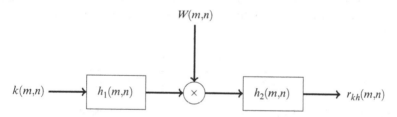

Fig. 5.6 Equivalent system representation of $D^{h^T} \Lambda_{hh}^{(k)} D^h$. The diagonal matrix Λ_{hh} gets reflected as the multiplication by $W(m, n)$ which makes the system shift variant

Using Eq. (5.43), the Fourier transform of $D^{h^T} \Lambda_{kh}^{(k)} D^h \underline{k}$ given by $\mathscr{R}_{kh}(\xi_x, \xi_y)$ is written as

$$\mathscr{R}_{kh}(\xi_x, \xi_y) = \mathscr{H}_2(\xi_x, \xi_y)[\mathscr{W}(\xi_x, \xi_y) \circledast (\mathscr{H}_1(\xi_x, \xi_y)\mathscr{K}(\xi_x, \xi_y))], \tag{5.44}$$

where $\mathscr{H}_1$ and $\mathscr{H}_2$, respectively, denote the DFT of h_1 and h_2 (all DFTs are of size $MN \times MN$), and $\mathscr{W}(\xi_x, \xi_y)$ is the Fourier transform of $W(m, n)$. Using Eqs. (5.41) and (5.44), for a general $\underline{k}$ and $\underline{x}$, the Fourier transform of Eq. (5.10) is written as

$$|\mathscr{X}^i|^2.\mathscr{K}^{i+1} + \mathscr{H}_2(\mathscr{W} \circledast (\mathscr{H}_1.\mathscr{K}^{i+1})) + \mathscr{H}_4(\mathscr{W} \circledast (\mathscr{H}_3.\mathscr{K}^{i+1})) = \overline{\mathscr{X}^{i+1}\mathscr{Y}}, \tag{5.45}$$

where $\mathscr{H}_3$ and $\mathscr{H}_4$ are the Fourier transforms of the masks h_3 and h_4, respectively.

From Eq. (5.45), due to the frequency domain convolutions in the second and third terms, one cannot write an expression for $\mathscr{K}^{i+1}$ as in Eq. (5.15) for smoothness regularizer. This happens since the system corresponding to Eq. (5.43), though linear, is not shift invariant. If the system were LSI, it would have been possible to write an expression for $\mathscr{K}^{(i+1)}$, which makes the analysis feasible. It can be seen that this system becomes linear shift invariant (LSI), if $W(m, n)$ is a constant. In order to make the analysis feasible, we assume that $W(m, n)$ is constant for each of the AM iterations and varies as the iteration changes. But the question arises as to what value should be chosen for the constant. The entries $W(m, n)$ at the $(i + 1)$th iteration are essentially the inverses of the magnitude of the gradient at each pixel at the ith iteration, and the AM iterations force the gradient to decrease after each iteration, i.e., the maximum value of the gradient decreases, leading to a decrement in the minimum entry of $W(m, n)$. Hence we take $W(m, n)$ as the constant matrix filled with the minimum value of $W(m, n)$ denoted by $\min(W^i(m, n))$, with i indicating the iteration. This modifies the regularization factor as $\lambda_k^i = \lambda_k \min(W^i(m, n))$. With this assumption, Eq. (5.45) can be rewritten as

$$\mathscr{K}^{i+1}(\xi_x, \xi_y) = \frac{\overline{\mathscr{X}^i(\xi_x, \xi_y)}\mathscr{Y}(\xi_x, \xi_y)}{|\mathscr{X}^i(\xi_x, \xi_y)|^2 + \lambda_k^i(k^i)\mathscr{R}(\xi_x, \xi_y)}, \tag{5.46}$$

where $\mathscr{R}$ is given in Eq. (5.17) and $\lambda_k^i(k^i)$ is a function of the PSF at the ith iteration, because of the way it is defined. Similarly, the image update equation in frequency domain is

$$\mathscr{X}^{i+1}(\xi_x, \xi_y) = \frac{\overline{\mathscr{K}^{i+1}(\xi_x, \xi_y)}\mathscr{Y}(\xi_x, \xi_y)}{|\mathscr{K}^{i+1}(\xi_x, \xi_y)|^2 + \lambda_x^i(x^i)\mathscr{R}(\xi_x, \xi_y)}, \tag{5.47}$$

where $\lambda_x^i(x^i)$ is defined similar to that of $\lambda_k^i(k^i)$. The expressions for the DFT of $\mathscr{K}$ and $\mathscr{X}$ as given by Eqs. (5.46) and (5.47) are similar to Eqs. (5.15) and (5.16) except for the difference that regularization terms $\lambda_x^i(x^i)$ and $\lambda_k^i(k^i)$ change with each AM iteration.

The phase of the Fourier transform of the reconstructed image is same as the phase of the observed image provided the iterations used the observed image as the initial value of the image. Under this condition the phase of the Fourier transform of the estimated PSF is zero. The proof is same as in [17] and is not reproduced here.

Let

$$u_{i+1} = |\mathscr{X}^{i+1}(\xi_x, \xi_y)|,$$

$$v_{i+1} = |\mathscr{K}^{i+1}(\xi_x, \xi_y)|,$$

and $r = \mathscr{R}(\xi_x, \xi_y)$, and $w = |\mathscr{Y}(\xi_x, \xi_y)|$. We use the modified notation $\lambda_x(x^i)$ and $\lambda_k(k^i)$ for the two regularization factors. Using these in the magnitude of Eqs. (5.46) and (5.47), we get

$$u_{i+1} = \frac{v_i w}{v_i^2 + \lambda_x(x^i)r}, \tag{5.48}$$

$$v_{i+1} = \frac{u_{i+1}w}{u_{i+1}^2 + \lambda_k(k^i)r}. \tag{5.49}$$

Substituting Eq. (5.49) in Eq. (5.48), the PSF update equation in the Fourier domain becomes

$$u_{i+1} = \frac{u_i w^2(u_i^2 + \lambda_k(k^i)r)}{u_i^2 w^2 + \lambda_x(x^i)r(u_i^2 + \lambda_k(k^i)r)^2}. \tag{5.50}$$

$$v_{i+1} = \frac{v_i w^2(v_i^2 + \lambda_x(x^i)r)}{v_i^2 w^2 + \lambda_k(k^i)r(v_i^2 + \lambda_x(x^i)r)^2}. \tag{5.51}$$

The magnitude spectrum updates of PSF and image at each iteration depends on the gradients in the spatial domain PSF and image, respectively, through the terms $\lambda_k(k^i)$ and $\lambda_x(x^i)$. Once the fixed point is reached, u and v do not change further and hence the regularization factors also remain constant. It maybe noted that this is the main difference between using the approximated TV and the smoothness regularizer of [17]. In [17], the regularization factors remain constant through out the iteration and the fixed point depends on the two regularization factors. In our analysis, both the regularization factors change in a signal (image/PSF) dependent fashion and reach a final value on which the fixed point depends.

From Eq. (5.50)

$$u_{i+1} = F(u_i), \tag{5.52}$$

where $F(u_i)$ is

$$F(u_i) = \frac{u_i w^2(u_i^2 + \lambda_k(k^i)r)}{u_i^2 w^2 + \lambda_x(x^i)r(u_i^2 + \lambda_k(k^i)r)^2}. \tag{5.53}$$

At the fixed point $F(u) = u$. Let λ_{kf} and λ_{xf} be the regularization factors corresponding to the PSF and the image, respectively, once the fixed points have been reached. When the fixed point is reached, the scenario is similar to that in [17], hence the proof is not reproduced here. Instead we highlight how the convergence points from our analysis differ from that of [17]. From Eq. (5.53) the fixed points (f_1, f_2) can be calculated as

$$u_1 = 0,$$

$$u_2 = \sqrt{\sqrt{\frac{\lambda_{kf}}{\lambda_{xf}}}w - \lambda_{kf}r}$$

For $w \le \sqrt{\lambda_{kf}\lambda_{xf}}\, r$, the fixed point u_1 is reached, and for $w > \sqrt{\lambda_{kf}\lambda_{xf}}\, r$, the fixed point u_2 is reached, provided the initialization for the image is non-zero. Hence the convergence points for the magnitude is given by a lemma similar to that in [17].

Lemma 5.2. *The magnitude spectra of the image and the PSF converge to*

$$M_T = \sqrt{\sqrt{\frac{\lambda_{kf}}{\lambda_{xf}}}|\mathscr{Y}| - \lambda_{kf}\mathscr{R}}$$

and $\sqrt{\dfrac{\lambda_{xf}}{\lambda_{kf}}}\, M_T$, *respectively, when* $|\mathscr{Y}|^2 > \lambda_{xf}\lambda_{kf}\mathscr{R}^2$, *else they both converge to zero.*

The difference between Lemmas 5.1 and 5.2 is that in Lemma 5.1 the convergence points can be obtained directly without using an iterative procedure whereas our analysis indicates convergence points that are reached once the iterations converge to the fixed point. But forcing the system to be space invariant makes the magnitude of the estimated image and the PSF proportional to each other. In the error analysis given in the next section we see that without this approximation, a space varying regularization is achieved.

Using the signal processing approach, given in Sect. 5.2.2, the convergence process can be explained as follows. At each iteration new values for the image and the PSF are obtained based on which the regularization parameters $\lambda_x^i(x^i)$ and $\lambda_k^i(k^i)$ are calculated. The next iteration proceeds with these modified regularization parameters. Hence the quadratic upper bounded TV behaves like an adaptive regularizer. From the discussion in Sect. 5.2.2, it is seen that the values of regularization parameters decide the quality of reproduced images. Since, in this case the regularization parameters are gradually adjusted to a value that depends on the total variation of the image and the PSF, convergence to a better solution is expected. It is seen from the experiments that the regularization parameters reduce and converge to a value determined by the TV of the image and PSF leading to an improved solution. This is expected since minimization of the TV norm leads to the reduction of the regularization factors at each iteration till the fixed point is reached, after which they remain constant. The product $\lambda_x^i\lambda_k^i$ is shown in Figs. 5.7d and 5.8d and the above behavior is observed in these plots.

The update equations Eqs. (5.46) and (5.47) are similar to that of the Wiener filter [166], except that this method is a purely deterministic approach. Instead of a fixed filter for restoring the image, the iterations converge to a filter which depends on the total variation of the image and the PSF. Hence the above approach may be thought of as a Wiener filter that adapts to the image and PSF total variations to give an optimum solution. It may also be noted that, the selection of the regularization factors shows a similarity to the step-size selection in gradient descent algorithms, where too low or too high initial value of regularization factors leads to convergence to trivial solutions. This again confirms with the results in Chap. 3.

5.3 Analysis of Error Due to Approximation

In Sect. 5.2.3, the $W(m, n)$ matrix was approximated to a constant matrix so that the resultant system is LSI at each iteration. This leads to convergence points for the image and the PSF which have magnitudes related to each other. The error due to this approximation is analyzed in this section. Keeping the PSF constant, the cost in Eq. (5.3) reduces to

$$C(\underline{x}) = \| \underline{y} - K\underline{x} \|^2 + \lambda_x \underline{x}^T D^T \Lambda_x D\underline{x}. \tag{5.54}$$

The estimated solution $\hat{\underline{x}}$ is the one which makes the gradient of Eq. (5.54) zero.

$$(K^T K + D^T \Lambda D)\hat{\underline{x}} = K^T \underline{y}. \tag{5.55}$$

The estimated solution is

$$\hat{\underline{x}} = (K^T K + D^T \Lambda D)^{-1} K^T \underline{y}. \tag{5.56}$$

The matrix inversion lemma (MIL)

$$(A + BCE)^{-1} = (I + A^{-1}BCE)^{-1} A^{-1}, \tag{5.57}$$

is used to expand the inverse term in Eq. (5.56). From Eq. (5.4) it is seen that Λ is a diagonal matrix with entries inversely proportional to the magnitude of the gradient at each pixel position. Λ is split into sum of two diagonal matrices

$$\Lambda = \Lambda_{min} + \Lambda_{res}, \tag{5.58}$$

where Λ_{min} is a diagonal matrix with all entries equal to the minimum non-zero valued entry of Λ, and Λ_{res} is a diagonal matrix with diagonal entry as the difference in diagonals of Λ and Λ_{min}. With this modification, $\hat{x}$ becomes

$$\hat{x} = (K^T K + D^T \Lambda_{min} D + D^T \Lambda_{res} D)^{-1} K^T \underline{y}. \tag{5.59}$$

Since all the diagonal elements in Λ_{min} have the same value (say λ_{min}), $D^T \Lambda_{min} D = \lambda_{min} D^T D$. Applying MIL to Eq. (5.59), with $A = K^T K + \lambda_{min} D^T D$, $B = D^T \Lambda_{res} D$, and $C = E = I$

$$\hat{x} = [I + (K^T K + \lambda_{min} D^T D)^{-1} D^T \Lambda_{res} D]^{-1} (K^T K + \lambda_{min} D^T D)^{-1} K^T \underline{y},$$

$$= [I + P]^{-1} \hat{x}_s, \tag{5.60}$$

where

$$P = (K^T K + \lambda_{min} D^T D)^{-1} D^T \Lambda_{res} D,$$

and

$$\hat{x}_s = (K^T K + \lambda_{min} D^T D)^{-1} K^T \underline{y}.$$

From the expression for $\hat{x}_s$, it is seen that this component corresponds to the solution obtained with a smoothness regularizer. If $|P| < 1$

$$(I + P)^{-1} = I - P + P^2 - P^3 + \dots \tag{5.61}$$

In the previous analysis, we approximated the solution at each iteration to the smoothness solution. Assuming that $|P| < 1$, the estimated solution can be written as

$$\hat{x} = \hat{x}_s + \sum_{i=1}^{\infty} (-1)^i P^i \hat{x}_s. \tag{5.62}$$

The operator P is a combination of two operators: $P_2 = D^T \Lambda_{res} D$ which extracts the high frequency component and $P_1 = (K^T K + \lambda_{min} D^T D)^{-1}$ which does a regularized inverse on its argument. To understand the effect of the operator P_2, we split it as

$$D^T \Lambda_{res} D = \sum_i D^T \Lambda_{min}^i D, \tag{5.63}$$

where Λ_{min}^i is a diagonal matrix with diagonal given by

$$\text{diag}(\Lambda_{min}^i) = \min(\text{diag}(\Lambda_{res}^{i-1})), \text{ filled at the non-zero locations of } \Lambda_{res}^{i-1}$$

$$\text{diag}(\Lambda_{res}^i) = \text{diag}(\Lambda_{res}^{i-1}) - \min(\text{diag}(\Lambda_{min}^i)) \text{ at the non-zero locations,} \tag{5.64}$$

where $\text{diag}(A)$ for a matrix A is defined as the diagonal of the matrix A. The $\min(.)$ function returns the smallest non-zero value of the vector argument. The number of diagonal elements becoming zero increases with i. Therefore

$$P_x \underline{x} = \sum_i D^T \Lambda_{min}^i D \underline{x} \tag{5.65}$$

$$= \sum_i \lambda_{min}^i \sum_j g_j, \tag{5.66}$$

where g_j is the negative of the gradient. So essentially P acting on the vector $\underline{x}_s$, has the effect of adding more components of higher frequency which are regularized by the P_1 operator. We note that this is a spatially varying regularization. The spatial points with small gradients get more regularization since the points which survive for larger values of i are those with small gradients and these points get regularized at each value of i (the elements of Λ_{res} are essentially the inverse of magnitude of gradient at each pixel point). This is reasonable since the noise amplification would be less for points of low gradients and hence it is feasible to take higher order derivatives of such points. Each time, the P operator is used, regularized high frequency components are added, with the Λ_{min}^i controlling noise amplification.

The condition $|P| < 1$ maps to the $\lambda_{min}(P_1)\lambda_{max}(P_2) \leq 1$, where $\lambda_{min}(P_1)$ is the minimum eigenvalue of P_1 and $\lambda_{max}(P_2)$ is the maximum eigenvalue of P_2. This provides a condition to select the image regularization factor. A similar analysis can be done for the PSF estimation part. This analysis gives an insight into the operation of the TV regularizer. The error incurred by the approximation discussed in Sect. 5.2.3 is that the high frequencies are not well regularized, leading to a loss of high frequency information, which is seen in Fig. 5.8b.

5.4 Demonstrations

In this section, we provide experimental results that substantiate our analysis. To estimate the image and the PSF, we need to solve Eqs. (5.9) and (5.10), respectively. Equation (5.9) is evaluated using the conjugate gradient descent method. Since K in Eq. (5.9) is the convolution matrix, $(K_i^T K_i + D^T \Lambda_{\underline{x}}^{(i)} D)\underline{x}$ is easily calculated through convolution using the masks h_1 and h_2 for evaluating the second term. We first attempt to estimate the PSF from Eq. (5.10) using both constrained and unconstrained methods, with constraints on k as mentioned in Eq. (5.6). For the constrained optimization implemented in Matlab, running even on a quad processor machine, the time taken for convergence and the memory requirements needed for specifying the constraints become too high to be of any practical use for large PSF size. For large PSF sizes the problem is converted to an unconstrained optimization by penalizing the cost function to enforce the PSF symmetry and by projecting the solution to ensure positivity and mean invariance. The results obtained are similar in both the cases, but the speed of convergence is much better for the conjugate gradient descent implementation.

For the constrained optimization case, the results obtained for a 3×3 PSF mask is shown in Fig. 5.7. In Fig. 5.7, the noise variance is zero, and the peak signal to noise ratio (PSNR) obtained in this case is 39.3 dB, which is a large improvement over the PSNR of the blurred input image (29 dB). The blurred and the deblurred images are shown in Fig. 5.7a, b. Figure 5.7c shows variation of the PSNR with iterations and Fig. 5.7d shows variations of the product of the regularization parameters with iterations. This product decreases initially and stabilizes at a value which depends on the total variation of the original image and the PSF. It is this nature of variation that leads to a better result when compared with the smoothness regularizer. The mean square error between the estimated and the original PSF is 4.54×10^{-6}.

The results of deblurring for a noise variance of 10 is given in Fig. 5.8. The mask used here is a Gaussian of size 5×5 with a spread of 3. The PSNR improvement obtained in this case is 3 dB. The mean square error in the estimated

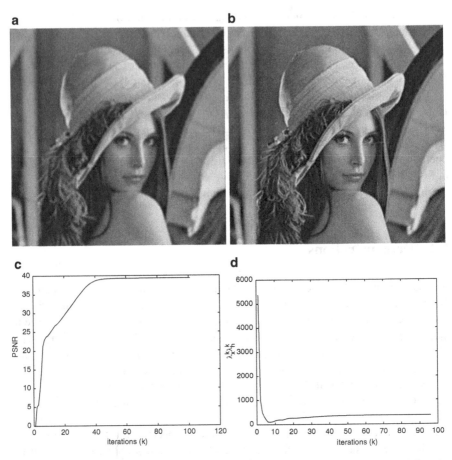

Fig. 5.7 Deconvolution for 3 × 3 Gaussian mask, with zero noise. (**a**) Blurred image. (**b**) Deconvolved image. (**c**) Variation of PSNR with iteration. (**d**) Variation of the product of regularization factors with iteration

PSF is 1.79×10^{-5}. Figure 5.8a shows the blurred and noisy image. Figure 5.8b shows the reconstructed image with approximation on $W(m, n)$, in which the ringing effect due to spectrum being forced to zero is visible. This may be compared with Fig. 5.8c, which shows the result of blind deconvolution with no approximation on $W(m, n)$. In the latter case no ringing is seen which confirms our hypothesis in Sect. 5.2.3. It is seen that in the case where the noise variance is non-zero, the estimated PSF could converge to three different values depending on the initial value of the PSF regularization term (λ_k). If λ_k is small, the PSF converges to impulse function and the estimated image is same as the original (Fig. 5.8d), which is as expected from the discussion in Sect. 5.2.2. If λ_k is large, the PSF converges to an averaging mask, with total variation close to zero, it reduces to a low pass filter.

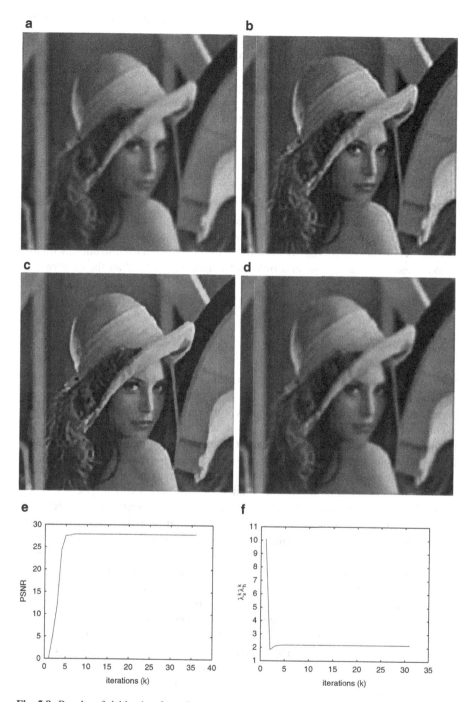

Fig. 5.8 Results of deblurring for a 5 × 5 mask, and noise variance 10. (**a**) Blurred and noisy input image. (**b**) Deblurring result with a constant approximation on $W(m, n)$. Ringing is noticed and PSNR is improved by 2 dB. (**c**) Deblurring result with quadratic upper bound on TV, the PSNR is improved by 3 dB. (**d**) Result of deblurring for low value of regularization parameter for PSF regularizer, a trivial solution is obtained. (**e**) Variation of PSNR with iteration for the method described. (**f**) Variation of the product of the two regularization terms with iteration

For an intermediate value of λ_k, which we determine empirically, the PSF converges to a value close to the original, with mean square error of the order of 10^{-5}.

In the implementation using conjugate gradient descent, while estimating the PSF it is not computationally economical to do an optimization over a $\underline{h}$ of size same as the image size, so the gradient with respect to h_i (ith pixel in the PSF), is evaluated by using

$$\| \underline{y} - H\underline{x} \|^2 = \sum_m \sum_n \left(y(m,n) - \sum_k \sum_l h(k,l)x(m-k,n-l) \right)^2. \quad (5.67)$$

This reduces the advantage of conjugate gradient descent in terms of speed, but still performs much faster than the constrained case. The result for conjugate gradient is comparable to that obtained using the constrained optimization, with the added advantage that it is several orders of magnitude faster than the constrained optimization case.

5.5 Discussions

We use the quadratic upper bounded TV function as regularizer for both the image and the PSF, which gives the advantage of quadratic cost functions in the AM iterations, making the problem amenable for convergence analysis. We provide a convergence analysis which shows that with each iteration of the alternate minimization algorithm the regularization factors get modified, eventually leading to an acceptable solution. We also provide an argument for why TV norm performs better than the smoothness regularizer. It is also shown that the convergence analysis provided in [17] could be derived in a much simpler way and is a special case of the analysis we have provided. In order to make Fourier domain analysis feasible we approximate the system to be LSI at each iteration. It is observed that the convergence point of the magnitude of the image and the PSF thus obtained are related to each other. We also do an analysis of the error introduced due to the approximation to make the system LSI at each iteration. It is shown that by making the system LSI, the regularization for higher spatial frequencies is lesser compared to the quadratic upper-bounded TV. From the analysis of the error term, it is seen that the approximation removes high frequency information leading to a quality degradation in the deconvolved result. The error analysis also shows that the TV regularizer essentially adds regularized high frequency components to the smoothness based solution, leading to a better solution than the smoothness based one. In the next chapter we look at a spatial domain convergence analysis.

Chapter 6
Spatial Domain Convergence Analysis

In the previous chapter we discussed the convergence analysis for blind deconvolution in the Fourier domain, using a modified TV norm as the regularizer. In this chapter we do a spatial domain convergence analysis with quadratic smoothness as the regularizer. As seen in Chap. 1, most formulations for blind restoration of an image involve minimization of a cost which is a function of two vectors, namely the image vector and the vector corresponding to the blur function. We also saw that alternate minimization (AM) is used in such cases to solve for the image and the PSF. Though alternate minimization is widely used in blind deconvolution, there is hardly any rigorous proof of convergence available. Csiszar et al. [27] reported a proof of convergence of AM algorithm applied to problems arising in information theory, which has a geometric flavor. The cost function considered in [27] is the Kullback-Leibler (KL) divergence [77], which is a measure of the difference between two probability distributions. Such a cost function arises in blind deconvolution when variational approach is used for estimating the PSF and the image [6, 92, 102, 106, 142, 143, 180]. In regularization or MAP based joint estimation the cost function is different as observed in the Chaps. 4 and 5. This necessitates extending the proof in [27] to analyze convergence of AM for blind deconvolution. Along similar lines to the proof in [27], Byrne [19] introduced a proof of convergence of AM problem applied to a limited class of problems. The type of problems addressed in [19] are those that do not involve a non-linear combination of the two unknown vectors. In [19] the convergence was proved using the three- and four-point properties which is simpler compared to the analysis in [27]. We use the three-point and four-point property to prove the convergence of AM for blind deconvolution with quadratic regularizers for both the image and the PSF.

The general form of the cost function for blind deconvolution using regularizers was given in Chaps. 4 and 5. It is repeated here for convenience with the symbols having the meanings as defined earlier. In the case of deterministic formulation we

© Springer International Publishing Switzerland 2014
S. Chaudhuri et al., *Blind Image Deconvolution: Methods and Convergence*,
DOI 10.1007/978-3-319-10485-0_6

had used $R_x(.)$ to denote the image regularizer, and for a random field we use $P_x(.)$. With an abuse of notation $R_x(x) = -\log(P_x(x))$.

$$C(\underline{x}, \underline{k}) = \| K\underline{x} - \underline{y} \|^2 - \lambda_x \log(P_x(\underline{x})) - \lambda_k \log(P_k(\underline{k})), \tag{6.1}$$

where $\underline{x}$ represents the column vector obtained from x by lexicographic ordering, $\underline{y}$, and $\underline{k}$ are defined similarly. K is the convolution matrix obtained from the PSF k. $P_x(.)$ and $P_k(.)$ are the priors, respectively, for x and k. λ_x and λ_k are the regularization factors for the image and the PSF, respectively.

Representing the image and the PSF as Gaussian MRF, $-\log(P_x(\underline{x}))$ can be written as

$$-\log(P_x(\underline{x})) = \sum_i \Delta_i^h(x)^2 + \Delta_i^v(x)^2, \tag{6.2}$$

where $\Delta_i^h(x)$ and $\Delta_i^v(x)$ represent the first order horizontal and vertical difference, respectively, at the ith pixel. Using matrix operators, Eq. (6.2) can be rewritten as

$$-\log(P_x(\underline{x})) = x^T D^T DX = \| Dx \|^2, \tag{6.3}$$

where $D = [D_h^T, D_v^T]^T$. D_h and D_v are matrices which when operating on a vector $\underline{x}$ generate the first order horizontal and vertical differences, respectively, of $\underline{x}$. Without any loss of generality, using a similar prior for PSF, the cost function can be modified as

$$C(\underline{x}, \underline{k}) = \| K\underline{x} - \underline{y} \|^2 + \lambda_x \| D_x \underline{x} \|^2 + \lambda_k \| D_k \underline{k} \|^2, \tag{6.4}$$

where D_x and D_k correspond to D defined for the image and the PSF, respectively. The only difference between D_x and D_k is in the dimension. D_x is of dimension $2MN \times MN$ for an $M \times N$ image and D_k has a dimension of $2PQ \times PQ$ for a PSF of size $P \times Q$. The image and the PSF are estimated by minimizing this cost function. Since $\underline{x}$ and $\underline{k}$ are both unknown variables, minimization is done in an alternate manner. The alternate minimization proceeds as follows. Let $(\hat{\underline{x}}^n, \hat{\underline{k}}^n)$ be the estimated image and the PSF at the nth iteration.

$$C(\hat{\underline{x}}^n, \hat{\underline{k}}^n) \xrightarrow[\text{keep } \hat{\underline{k}}^n \text{ fixed}]{\text{minimize}} C(\hat{\underline{x}}^{n+1}, \hat{\underline{k}}^n) \xrightarrow[\text{keep } \hat{\underline{x}}^{n+1} \text{ fixed}]{\text{minimize}} C(\hat{\underline{x}}^{n+1}, \hat{\underline{k}}^{n+1}). \tag{6.5}$$

Since $C(\hat{\underline{x}}^{n+1}, \hat{\underline{k}}^{n+1}) \le C(\hat{\underline{x}}^{n+1}, \hat{\underline{k}}^n) \le C(\hat{\underline{x}}^n, \hat{\underline{k}}^n)$, the sequence $\{C(\hat{\underline{x}}^k, \hat{\underline{k}}^k)\}$ is a decreasing sequence. A frequency domain approach to prove the convergence for this problem is provided in [17]. We use the traditional three-point four-point property [19,27] based convergence analysis of alternate minimization algorithm to prove the convergence in the blind deconvolution case. A brief overview of the proof by Byrne [19] is given in Sect. 6.1. In Sect. 6.2, the three-point property is proved for the blind deconvolution problem. The conditions under which the four-point property holds are derived in the Sect. 6.3 followed by discussions in Sect. 6.4.

6.1   Three- and Four-Point Property Based Convergence Analysis

In the previous section, we saw that AM leads to iterative optimization of the cost given by Eq. (6.4). The three- and four-point analysis in [19] shows that provided the two variables involved in the AM satisfy the three- and four-point property then the algorithm converges to the infimum of the cost function. Let $\mathbf{X}$ and $\mathbf{K}$ be arbitrary sets and C is the function defined as

$$C : \mathbf{X} \times \mathbf{K} \to \mathbb{R}. \tag{6.6}$$

The function C satisfies $-\infty < C(x, k) < \infty$, $x \in \mathbf{X}$, $k \in \mathbf{K}$. Let e be the infimum of C, and let $(\hat{x}, \hat{k})$ be the point where the function reaches the value e, i.e., $C(\hat{x}, \hat{k}) = e$. From Eq. (6.5), it is seen that the sequence $\{C(x^n, k^n)\}$ is a decreasing one, which is illustrated below.

$$C(x^n, k^n) \xrightarrow[\text{keep } k^n \text{ fixed}]{\text{minimize}} C(x^{n+1}, k^n) \xrightarrow[\text{keep } x^{n+1} \text{ fixed}]{\text{minimize}} C(x^{n+1}, k^{n+1}). \tag{6.7}$$

This sequence is bounded below by e and since it is decreasing it converges to a value $E \geq e$. The sufficient conditions for the sequence $\{C(x^n, k^n)\}$ to converge to e are that the point x must satisfy the three-point and four-point properties, for any x for which there is a k with $C(x^n, k^n) \geq C(x, k)$ for all n.

6.1.1   Three- and Four-Point Properties

A non-negative function (d) is required for defining the three- and four-point properties. Let $d : \mathbf{X} \times \mathbf{X} \to \mathbb{R}^+$ be the non-negative function satisfying $d(x, x) = 0$ for all $x \in \mathbf{X}$. The three- and four-point properties are defined for a point $x \in \mathbf{X}$.

Definition 6.1. A point $x \in \mathbf{X}$ has the three-point property, if

$$C(x, k^n) \geq C(x^{n+1}, k^n) + d(x, x^{n+1}) \qquad n = 1, 2, \ldots \tag{6.8}$$

From Eq. (6.7), $C(x^n, k^n) \geq C(x^{n+1}, k^n)$. This means that if x satisfies the three-point property, then for a fixed k^n, the cost at any point (x, k^n) is larger than the cost at (x^{n+1}, k^n) by at least $d(x, x^{n+1})$. If the point x^n satisfies the 3-point property, then Eq. (6.8) becomes

$$C(x^n, k^n) - C(x^{n+1}, k^n) \geq d(x^n, x^{n+1}), \tag{6.9}$$

which defines a lower bound for the decrease in the cost function as a result of updating the first variable.

Definition 6.2. A point $x \in X$ has the four-point property, if

$$d(x, x^n) + C(x, k) \geq C(x, k^n), \qquad \forall k \in K \text{ and } \forall n \tag{6.10}$$

Since this definition is true for all k and n, it should be true for some $k = k^m$, i.e.,

$$C(x, k^n) - C(x, k^m) \leq d(x, x^n) \tag{6.11}$$

for a fixed n. For $m < n$ and no change in x, Eq. (6.11) is true due to Eq. (6.7). If $m > n$, we are looking at the point k^m which was obtained by minimizing the function at k^n. According to four-point property, if x has this property then even for $m > n$ the sequence $\{C(x, k^n)\}$ is such that the difference between the terms for some $m > n$ is always less than $d(x, x^n)$. If the point x^n satisfies the four-point property, then Eq. (6.10) becomes

$$C(x^n, k^{n-1}) - C(x^n, k^n) \leq d(x^n, x^{n-1}), \tag{6.12}$$

which gives an upper bound to the amount by which the function decreases due to the updation of second variable. We provide a graphical illustration (Fig. 6.1) of the three- and four-point properties for the case when the point x^n satisfies both these properties. In Fig. 6.1, starting from the point (x^n, k^{n-1}), updating k decreases the cost by an amount which is at the most $d(x^n, x^{n-1})$, i.e., if δC_k is the amount by which $C(x, k)$ decreases when k is updated, then $\delta C_k \in [0, d(x^n, x^{n-1})]$. This follows from the four-point property. Now, from the point (x^n, k^n), if x is updated, the new point reached is (x^{n+1}, k^n) and the amount by which the cost decreases $\delta C_x \geq d(x^n, x^{n+1})$ which follows from the three-point property, i.e., the cost decreases at least by $d(x^n, x^{n+1})$.

Suppose that the cost E at the point of convergence is greater than the infimum e. Then there exists a point $(\tilde{x}, \tilde{k})$ such that

$$E > C(\tilde{x}, \tilde{k}) \geq e. \tag{6.13}$$

Assume that we start with a k^0 such that $C(x, k^0)$ is finite for all x. If the point $\tilde{x}$ satisfies the three- and four-point properties, then

$$d(\tilde{x}, x^1) \leq C(\tilde{x}, k^0) - C(x^1, k^0), \tag{6.14}$$

$$C(\tilde{x}, k^1) \leq d(\tilde{x}, x^1) + C(\tilde{x}, k^0). \tag{6.15}$$

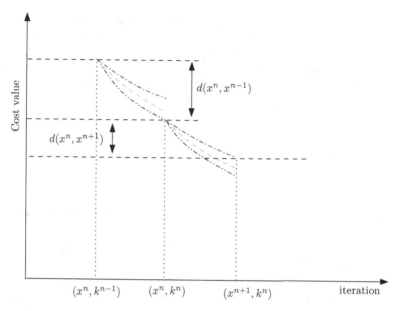

Fig. 6.1 Illustration of the implication of x^n satisfying the three- and four-point properties; the -.-, - -, and -..-, lines indicate the possible paths followed by the cost function due to the minimization process. The decrement is bound to lie within $[0, d(x^n, k^{n-1})]$ while k is updated, and is larger than $d(x^n, x^{n+1})$ when x is updated

From Eq. (6.14) $d(\tilde{x}, x^1)$ is finite and from Eq. (6.15) $C(\tilde{x}, k^1)$ is finite. By repeated application of Eqs. (6.14) and (6.15), for varying n, it is seen that $C(\tilde{x}, k^n)$ and $d(\tilde{x}, x^n)$ are finite for all n. The theorem which states the conditions for convergence of $C(x^n, k^n)$ is given in [19] as

Theorem 6.1. *Suppose that x has the three- and four-point properties, for any x for which there is a k with $C(x^n, k^n) \geq C(x, k)$ for all n, then $E = e$ and so*

$$C(x^n, k^n) \to e. \tag{6.16}$$

Proof. Assume that $E > e$ and there exists $(\tilde{x}, \tilde{k})$ as mentioned in Eq. (6.13). Since $(\tilde{x}, \tilde{k})$ satisfies both the three- and four-point properties

$$C(\tilde{x}, \tilde{k}) \geq C(x^{n+1}, k^n) + d(\tilde{x}, x^{n+1}),$$
$$d(\tilde{x}, x^n) + C(\tilde{x}, \tilde{k}) \geq C(\tilde{x}, k^n). \tag{6.17}$$

Combining these two equations in Eq. (6.17), we get

$$d(\tilde{x}, x^n) - d(\tilde{x}, x^{n+1}) \geq C(x^{n+1}, k^n) - C(\tilde{x}, \tilde{k}). \tag{6.18}$$

Since $C(x^{n+1}, k^n) > C(\tilde{x}, \tilde{k})$, from Eq. (6.18), it follows that $d(\hat{x}, x^n)$ is a decreasing sequence and we had shown earlier that $d(\hat{x}, x^n)$ is finite for each n. Hence it follows that the sequence $\{d(\hat{x}, x^n)\}$ converges to zero. This means that $C(x^{n+1}, k^n)$ converges to $C(\tilde{x}, \tilde{k})$, which is a contradiction since we started with the assumption that the cost E at the point of convergence of $\{C(x^{n+1}, k^n)\}$ is greater than $C(\tilde{x}, \tilde{k})$. Hence $E = e$.   □

In the next section, we proceed with proof of convergence for blind deconvolution using AM by proving the three-point property for all points in the set $\mathbf{X}$ and derive the conditions under which the four-point property is satisfied.

6.2 Proof of Three-Point Property for Blind Deconvolution

The cost function that is to be minimized is Eq. (6.4). The set $\mathbf{X}$ is the set of all lexicographically ordered images of size $M \times N$ and $\mathbf{K}$ is the set of all lexicographically ordered PSFs of size $P \times Q$, $\underline{x} \in \mathbf{X}$, and $\underline{k} \in \mathbf{K}$. For convenience we represent $\underline{x}$ by x, $\underline{y}$ by y, and $\underline{k}$ by k. With this change of notation, the cost function can be rewritten as

$$C(x, k) = \| Kx - y \|^2 + \lambda_x \| Dx \|^2 + \lambda_k \| Dk \|^2 \tag{6.19}$$

Since 3-point property describes the behavior of the optimization, while fixing the second variable, we fix k to get a cost which is a function of the variable x alone.

$$C_1(x) = \| Kx - y \|^2 + \lambda_x \| Dx \|^2 . \tag{6.20}$$

To prove the 3-point property it needs to be shown that

$$C(x, k^n) - C(x^{n+1}, k^n) \geq d(x, x^{n+1}). \tag{6.21}$$

Assume that we have a starting vector $k^0 \in \mathbf{K}$ such that $C(x, k^0)$ is finite for all $x \in \mathbf{X}$. This is true because selecting a discrete impulse as the initial PSF, satisfies the above condition. Let x^1 minimize $C(x, k^0)$ over all $x \in \mathbf{X}$. Since x^1 minimizes $C(x, k^0)$, it is a stationary point of the function. i.e.,

$$\frac{\partial}{\partial x} C(x, k^0)\big|_{x=x^1} = 0 \tag{6.22}$$

From Eqs. (6.20) and (6.22)

$$(K^{0^T} K^0 + \lambda_x D^T D)x^1 - K^{0^T} y = 0, \tag{6.23}$$

where K^0 is the convolution matrix corresponding to the PSF k^0. The function $d(.)$ in Eq. (6.21) is not identified yet, and we arrive at the function in the process of proving the inequality in Eq. (6.21). Evaluating the L.H.S. of Eq. (6.21)

$$C(x, k^0) - C(x^1, k^0) = x^T A x - x^{1^T} A x^1 - 2y^T K^0 (x - x^1), \qquad (6.24)$$

where the matrix A is defined as $A = K^{0^T} K^0 + \lambda_x D^T D$.

Replacing $y^T K^0 (x - x^1)$ in Eq. (6.24) by Eq. (6.23), gives

$$C(x, k^0) - C(x^1, k^0) = x^T A x - x^{1^T} A x^1 - 2(x - x^1)^T A x^1$$

$$= x^T A x - 2x^T A x^1 + x^{1^T} A x^1. \qquad (6.25)$$

Since A is a real symmetric matrix, it can be decomposed as $A = U H U^T$, where U is an orthogonal matrix and H is a diagonal matrix whose entries are the eigenvalues of A. Using this decomposition in the place of A in Eq. (6.25), and defining vectors, $U^T x = \underline{a}$ and $U^T x^1 = \underline{b}$, gives

$$C(x, k^0) - C(x^1, k^0) = x^T U H U^T x - 2x^T U H U^T x^1 + x^{1^T} U H U^T x^1,$$

$$= \underline{a}^T H \underline{a} - 2\underline{a}^T H \underline{b} + \underline{b}^T H \underline{b}, \qquad (6.26)$$

$$= \sum_i \eta_i (a_i - b_i)^2, \qquad (6.27)$$

$$\geq \eta_{min} \sum_i (a_i - b_i)^2, \qquad (6.28)$$

where η_i's are the eigenvalues of A, and η_{min} is the smallest eigenvalue of A, and a_i, b_i are the ith elements of $\underline{a}$ and $\underline{b}$, respectively. Since $a_i = (U^T \underline{x})_i$ and $b_i = (U^T \underline{x}^1)_i$,

$$\sum_i (a_i - b_i)^2 = (x - x^1)^T U U^T (x - x^1),$$

$$= (x - x^1)^T (x - x^1),$$

$$= \| x - x^1 \|^2. \qquad (6.29)$$

Combining Eqs. (6.29) and (6.28)

$$C(x, k^0) - C(x^1, k^0) \geq \eta_{min} \| x - x^1 \|^2. \qquad (6.30)$$

This inequality is of the form Eq. (6.8) with the non-negative function d defined as

$$d(x, x^1) = \eta_{min} \| x - x^1 \|^2.$$

But the problem in taking it as the definition of $d(.)$ is that, since η_{min} is the smallest eigenvalue of A which depends on K, it keeps changing with iteration since K is updated in each iteration of the AM algorithm. But we need a function d which does not change with iterations. In order to obtain such a function we find a lower bound for η_{min}, the lowest eigenvalue of A.

6.2.1 Lower Bound for the Smallest Eigenvalue

In this sub-section we analyze the variation of η_{min} as the convolution matrix K changes, and also derive an approximate expression for η_{min} when the PSF is a Gaussian blur. From Eq. (6.24) A is of the form

$$A = K^T K + \lambda_x (D_h^T D_h + D_v^T D_v), \qquad (6.31)$$

where D_h and D_v are as described in Eq. (6.3). It may be noted that K, D_h, and D_v are convolution matrices associated with the PSF, horizontal derivative and vertical derivative, respectively. This means that all the three matrices are block circulant with circulant blocks (BCCB) and hence are diagonalized by the discrete Fourier transform (DFT). The diagonal elements correspond to the DFT of the first column of the BCCB matrix. Let F_M be the DFT matrix of size $M \times M$ and F_N, the DFT matrix of size $N \times N$. The DFT matrix $F = F_M \otimes F_N$ ($\otimes$ denotes the Kronecker product), which is of size $MN \times MN$ diagonalizes the above BCCB matrices.

Since all three matrices operate on the same image data, the circulant blocks are of same size. Hence A is also a BCCB matrix which is diagonalized by F. This leads to

$$H = F^* A F,$$
$$= F^* (K^T K + \lambda_x D_h^T D_h + \lambda_x D_v^T D_v) F$$
$$= |H_k|^2 + \lambda_x (|H_{hd}|^2 + |H_{vd}|^2), \qquad (6.32)$$

where $*$ represents conjugate transpose operation, $|H_k|^2$, $|H_{hd}|^2$, and $|H_{vd}|^2$, respectively, represent the magnitude squared responses of the blur, the horizontal derivative and the vertical derivative. Hence the eigenvalues of A are the sum of magnitude square of the DFT of a low pass and two high pass filters. Figure 6.2a has the minimum eigenvalue of A plotted as a function of the PSF size P ($P = Q$ was chosen), for a Gaussian blur with spread chosen as $0.6P$. The size of the convolution matrix K is $MN \times MN$, where $M \times N$ is the size of the image. For lower values of the PSF size (P), K is mostly sparse and as P approaches M, the rows of K becomes approximately linear combinations of adjacent rows [53] making the smallest eigenvalues tend towards zero which pulls down the eigenvalue of A. This tendency is seen in Fig. 6.2a where $M = N = 80$ was chosen. Fig. 6.2b shows the

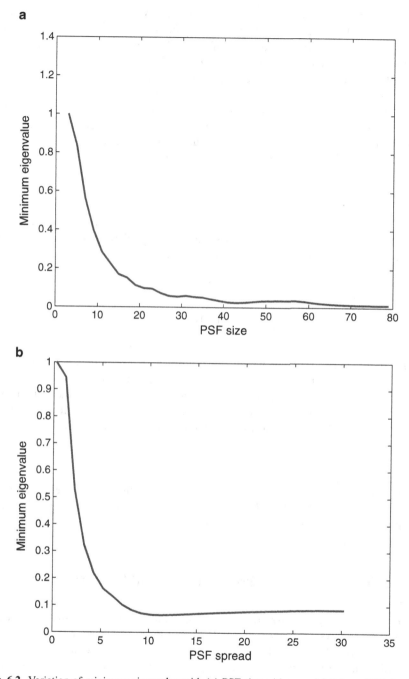

Fig. 6.2 Variation of minimum eigenvalue with (**a**) PSF size with spread 0.6 times PSF size (**b**) PSF spread for a PSF size of 31

variation of the smallest eigenvalue of A as a function of the spread, keeping the size of the PSF fixed. Fixing the size and increasing the spread makes the resultant PSF to tend towards a uniform PSF because of which the minimum eigenvalue becomes more or less a constant after some value of spread, which is observed in the plot. These plots indicate that the smallest eigenvalue is a function of the PSF size for a given spread and for a given PSF it is a function of the spread.

We derive a closed form expression for the minimum eigenvalue for the continuous domain equivalent for the simpler case of a one dimensional signal and Gaussian blur. Under this assumption, the sum of the squared magnitude response of the derivative filter and the low pass filter can be written as

$$G(\omega) = \omega^2 + e^{-\sigma^2 \omega^2}, \tag{6.33}$$

where ω is the continuous domain frequency variable, and σ the spread of the Gaussian blur. The first term in Eq. (6.33) is due to the derivative operation (high pass filter) and the second term due to the Gaussian low pass filter. The minimum of $G(\omega)$ occurs at $\omega^2 = \dfrac{\ln \sigma^2}{\sigma^2}$, and has value

$$\eta G_{min} = \frac{1 + \ln \sigma^2}{\sigma^2} = f(\sigma), \tag{6.34}$$

where ηG_{min} is the point at which the minimum of $G(\omega)$ occurs. This shows that η_{min} can be expressed a function of the blur parameter σ, even though K, the convolution matrix keeps varying with iteration. Here G is the continuous domain equivalent of the discrete operator A and ηG_{min} is equivalent to the smallest eigenvalue of A. This enables us to write, the non-negative function $d(.)$ as

$$d(x, x^1) = f(\sigma) \parallel x - x^1 \parallel^2, \tag{6.35}$$

where $f(\sigma)$ is a function of the blur parameter σ. Since the PSF size remains fixed, we treat η_{min} as a function of the spread of PSF alone. For other type of blurs, a lower bound of η_{min} could be found and used in place of $f(\sigma)$.

6.3 Proof of Four-Point Property for Blind Deconvolution

The proof for four-point property proceeds on similar lines as for the three-point property. For analysis purposes, we make use of the fact that the convolution Kx is a circular convolution and hence $Kx = Xk$, where X is the convolution matrix obtained from x. Four-point property analyzes the cost function behavior when it is minimized w.r.t. k by keeping x a constant. The cost function in this case is

$$C_2(k) = \parallel Xk - y \parallel^2 + \lambda_k \parallel Dk \parallel^2. \tag{6.36}$$

It follows from Eqs. (6.10) and (6.35) that, to prove the four-point property it has to be shown that

$$f(\sigma) \parallel x - x^1 \parallel^2 + C(x,k) - C(x,k^1) \geq 0. \qquad (6.37)$$

By interchanging the role of x and k in the analysis Eq. (6.24) in Sect. 6.2, we get

$$C(x,k) - C(x,k^1) = k^T B k - k^{1^T} B k^1 - 2y^T X(k - k^1), \qquad (6.38)$$

where B is defined as $B = X^T X + \lambda_k D^T D$. The right hand side of Eq. (6.38) can be written as

$$k^T B k - k^{1^T} B k^1 = (k - k^1)^T B(k - k^1) - 2k^{1^T} B k^1 + 2k^{1^T} B k,$$

$$= (k - k^1)^T B(k - k^1) + 2k^{1^T} B(k - k^1). \qquad (6.39)$$

The real and symmetric matrix B can be decomposed as $B = V \Gamma V^T$, where Γ is a diagonal matrix whose entries are the eigenvalues of B. This leads to

$$k^T B k - k^{1^T} B k^1 = (k - k^1)^T V \Gamma V^T (k - k^1) + 2k^{1^T} B(k - k^1). \qquad (6.40)$$

With some algebraic manipulations, Eq. (6.40) can be modified and used in Eq. (6.38) to obtain

$$C(x,k) - C(x,k^1) = \sum_i \gamma_i (c_i - d_i)^2 + 2(k - k^1)^T (Bk^1 - X^T y), \qquad (6.41)$$

where, γ_is are the eigenvalues of B. c_i and d_i denote the ith element of the vectors $\underline{c}$, $\underline{d}$, respectively, which are defined as $\underline{c} = V^T k$, and $\underline{d} = V^T k^1$. V is the orthogonal matrix which appears in the decomposition of B. Substituting Eq. (6.41) in Eq. (6.37) we get

$$f(\sigma) \parallel x - x^1 \parallel^2 + C(x,k) - C(x,k^1) \geq f(\sigma) \parallel x - x^1 \parallel^2 +$$
$$\gamma_{min} \parallel k - k^1 \parallel^2 + 2(k - k^1)^T (Bk^1 - X^T y), \qquad (6.42)$$

where γ_{min} is the minimum eigenvalue of B. Now, to show that the four-point property holds it remains to show that the right side of the inequality Eq. (6.42) is greater than or equal to zero, i.e., to show that

$$f(\sigma) \parallel x - x^1 \parallel^2 + \gamma_{min} \parallel k - k^1 \parallel^2 + 2(k - k^1)^T (Bk^1 - X^T y) \geq 0. \qquad (6.43)$$

Using z to denote the vector $Bk^1 - X^T y$, Eq. (6.43) becomes

$$f(\sigma) \parallel x - x^1 \parallel^2 + \gamma_{min} \parallel k - k^1 \parallel^2 + 2(k - k^1)^T z \geq 0. \qquad (6.44)$$

To see what the vector z corresponds to, we simplify the term $Bk^1 - X^T y$. Let $X = X^1 + \Delta X$, where X^1 is the convolution matrix corresponding to x^1.

$$Bk^1 - X^T y = Bk^1 - (X^{1^T} + \Delta X^T) y$$
$$= Bk^1 - X^{1^T} y - \Delta X^T y \tag{6.45}$$

Since k^1 minimizes Eq. (6.36) with $x = x^1$, k^1 is a stationary point of the cost, giving (similar to Eq. (6.23))

$$(X^{1^T} X^1 + \lambda_k D^T D)k^1 - X^{1^T} y = 0. \tag{6.46}$$

Substituting for $X^{1^T} y$ from Eq. (6.46) and using the definition of B, Eq. (6.45) gets modified as

$$Bk^1 - X^T y = 2X^{1^T} \Delta X k^1 + \Delta X^T (\Delta X k^1 - y). \tag{6.47}$$

The fact that X and ΔX are BCCB has been made use in deriving Eq. (6.47). Since these two matrices are BCCB, the matrix product becomes commutative [31]. From Eq. (6.47) it is seen that the vector $z = Bk^1 - X^T y$ is an element of $\mathbf{K}$ and is defined in terms of the difference between the point for which four-point property is defined and x^1.

For the four-point property to be satisfied, the inequality in Eq. (6.44) should be satisfied. The last term of Eq. (6.44) is the inner product of the vectors $(k - k^1)$ and z which can be written as

$$(k - k^1)^T z = \| k - k^1 \| \| z \| \cos \theta, \tag{6.48}$$

where θ is the angle between the vectors, $k - k^1$ and z. Using Eq. (6.48) in Eq. (6.44)

$$f(\sigma) \| x - x^1 \|^2 + \gamma_{min} \| k - k^1 \|^2 + 2 \| k - k^1 \| \| z \| \cos \theta \geq 0. \tag{6.49}$$

For the quantity in Eq. (6.49) to be greater than or equal to zero, the following condition should be satisfied

$$\cos \theta \geq -\frac{f(\sigma) \| x - x^1 \|^2 + \gamma_{min} \| k - k^1 \|^2}{2 \| k - k^1 \| \| z \|}. \tag{6.50}$$

The R.H.S. of this inequality is always negative and the inequality is always satisfied if

$$\frac{f(\sigma) \| x - x^1 \|^2 + \gamma_{min} \| k - k^1 \|^2}{2 \| k - k^1 \| \| z \|} \geq 1. \tag{6.51}$$

The condition in Eq. (6.51) is satisfied provided the step-size of the image update ($\| x - x^1 \|$) is larger than $\sqrt{\| k - k^1 \|} \| z \|$. Though we use the term step-size, it is actually the distance of the point x which satisfies the three- and four-point properties from the update x^1. The quantity z also indicates an update in the blur, so this requirement may be understood as the step-size of image update should be larger than step-size of PSF update. In addition the minimum eigenvalue of the sum of blur convolution matrix and regularization matrix should not be too low, which indicates that the regularizer should be able to improve the condition number of the blur convolution matrix. In this case all the points in $\mathbf{X}$ satisfy the three- and four-point properties and the algorithm converges to the infimum. If the inequality in Eq. (6.51) is not satisfied then there would be an angular region around the angle $180°$ where the inequality Eq. (6.50) would fail leading to points where the four-point property is not satisfied. Let $\mathbf{P}$ be the set of all points in $\mathbf{X}$ which satisfy Eq. (6.50) when Eq. (6.51) fails. Since every point in $\mathbf{X}$ satisfies the three-point property, all points in $\mathbf{P}$ satisfy both the three- and four-point properties. The alternate minimization process converges only if the implementation method ensures that at each iteration $\hat{x} \in \mathbf{P}$, where $\hat{x}$ is the estimate at each iteration.

6.4 Discussions

Alternate minimization gives rise to a sequence of image and PSF pairs which decreases the cost at each step and if all the points in the space $\mathbf{X}$ satisfies both the three- and four-point properties, then the above sequence converges to the point where the cost function reaches its infimum. We prove the three-point property for the blind deconvolution problem for all the points in $\mathbf{X}$. In order to prove the three- and four-point property for blind deconvolution, the non-negative function d needs to be obtained. We obtain the non-negative function and show that it depends on the lower bound of the smallest eigenvalue of a convolution matrix which arises from the blur and the regularizer. For the special case of a Gaussian blur, we show that η_{min} is a function of the spread of the Gaussian blur and for the corresponding continuous case we derive the closed form expression for the lower bound. For the case of four-point property, we show that if the step size of the image update is sufficiently large compared to that of PSF update and if the regularization is sufficiently strong, the convergence to the infimum is ensured. If the step size is not chosen appropriately, there are points in $\mathbf{X}$ where the four-point property is not satisfied. In this case convergence is guaranteed only if the implementation methodology makes sure that the image estimates at each iteration falls in the subset of $\mathbf{X}$ in which both the properties are satisfied.

Chapter 7
Sparsity-Based Blind Deconvolution

In this chapter we explore sparsity based regularizers for solving the blind deconvolution problem. In Chap. 4, we observed that joint MAP gives a trivial solution if the PSF regularizer is not chosen appropriately. The tendency of MAP to provide a trivial solution was corrected by choosing a proper PSF regularization term and the regularization factor. The main reason for MAP to fail in the absence of an appropriate PSF regularizer is that the cost due to the image regularization term decreases with an increase in blur amount. This indicates that we should explore another solution to overcome the failure of MAP by looking for a regularizer for which the cost increases with the blur amount. One such regularizer was proposed by Krishnan et al. [76], which works in the sparse derivative domain. But this regularizer has the disadvantage that the cost increases with blur only for certain images. Based on our analysis of the behavior of the cost for different types of images we classify the images into two major types – Type 1 and Type 2, with two subclasses for the Type 2 image (2a and 2b), the details of which we give in subsequent sections. The prior proposed in [76] works well for Type 1 images. For Type 2a images, the cost in [76] does not exhibit monotonically rising behavior for low blur amounts, and for Type 2b images, the cost actually decreases with blur amount making the regularizer ineffective.

We show that the regularizer proposed in [76] does not have a monotonically rising cost for all images, and define a class of regularizers which exhibit this behavior for all types of images. In addition we also look at an alternate sparse domain, the wavelet domain, and suggest regularizers in this domain which have the property that cost increases with blur size, thus ensuring that the cost attains its minimum value for the original sharp image.

In Sect. 7.1 we motivate the choice of the sparse domain and the regularizer, Sect. 7.2 discusses the deconvolution process. Our findings are reported in Sect. 7.3, with discussions in Sect. 7.4.

© Springer International Publishing Switzerland 2014
S. Chaudhuri et al., *Blind Image Deconvolution: Methods and Convergence*,
DOI 10.1007/978-3-319-10485-0_7

7.1 Choice of Regularizer

In Chap. 4, the general form of the commonly used image prior [2] was given and is repeated here for convenience.

$$\log p(x) = -\sum_{i}(|g_{h,i}(x)|^{\alpha} + |g_{v,i}(x)|^{\alpha}) + C. \tag{7.1}$$

It may be noted that, in general, the image prior is a function of the gradients $(g_{h,i}(x), g_{v,i}(x))$ since the solutions are restricted to be smooth. The component added to the overall cost function by the image prior is

$$R_{x}(x) = -\log p(x) = \sum_{i}(|g_{h,i}(x)|^{\alpha} + |g_{v,i}(x)|^{\alpha}).$$

It is easy to see why $R_{x}(x)$ prefers a blurred solution over the sharp image; the gradients of the blurred image being of lower magnitude than that of the sharp one, the cost in the blurred case goes down. This behavior is seen in Fig. 7.1 which plots $R_{x}(x)$ as a function of the blur size. The norms are normalized w.r.t. the corresponding norm of the original sharp image. The PSF used to obtain the results is Fig. 7.1 is Gaussian with size of the PSF as 6σ where σ is the spread. It is seen from the figure that for both the ℓ_1 and ℓ_2 norms, the cost of the regularizer goes down as the amount of blur increases for all the three images making both the norms unsuitable as regularizers. It may be noted from Fig. 7.1 that for the first image (Fig. 7.1a), the normalized value of ℓ_1 norm is larger than the corresponding ℓ_2 norm, whereas both have almost same value for the image in Fig. 7.1c and the normalized ℓ_2 norm has a larger value compared to the corresponding ℓ_1 norm for the last image in Fig. 7.1e. We will see in later sections that not only the absolute value of the norms is important, but also the rate at which the two norms reduce as a function of the blur amount is important.

7.1.1 ℓ_1/ℓ_2 Norm and Its Disadvantages

One way of overcoming the disadvantage of the regularizer in Eq. (7.1) is by selecting a function which has a cost that increases with the amount of blur. One such regularizer was proposed by Krishnan et al. [76] as the ℓ_1/ℓ_2 norm of the image derivative.

$$R_{s}(x) = \frac{\ell_1(Dx)}{\ell_2(Dx)}, \tag{7.2}$$

where $R_{s}(x)$ is the sparsity based regularizer, and Dx is the first order difference of the image x. It is shown in [76] that $\ell_1(Dx)/\ell_2(Dx)$ is a valid sparsity measure, which is scale invariant. Since the image and the PSF are estimated using

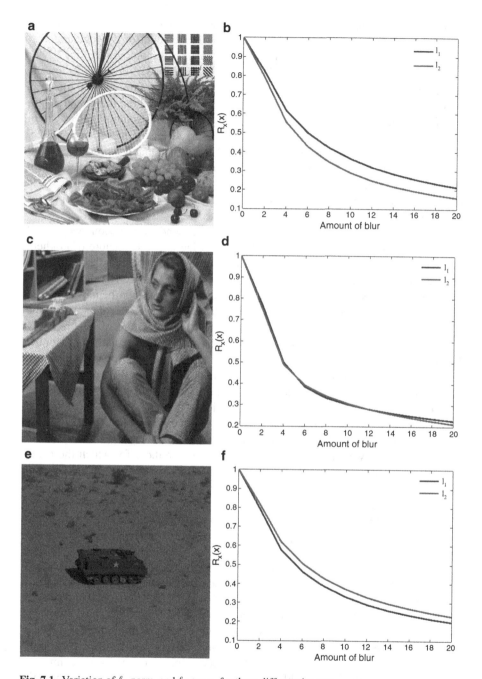

Fig. 7.1 Variation of ℓ_1 norm, and ℓ_2 norm, for three different images

an iterative procedure, during the estimation process, the argument of the image regularizer is a blurred and noisy image. To see the behavior of the regularizer as a function of blur, we plot $R_s(y)$ as a function of the blur amount in Fig. 7.2 for the same set of images in Fig. 7.1. The ℓ_1 and ℓ_2 norms used in Eq. (7.2) are the normalized ones, normalized w.r.t. the corresponding norms of the original image x. It is seen that the behavior of $R_s(y)$ is not the same for the three images shown in Fig. 7.1. It can be observed from Fig. 7.2a that $R_s(y)$ increases monotonically with the amount of blur. This monotonicity is not observed for the image in Fig. 7.2b, also in this case the range of variation of $R_s(y)$ is so small that it can be considered to be a constant. Fig. 7.2c shows that for the third image shown in Fig. 7.1 the norm proposed in [76] has a cost which decreases with the amount of blur. From these observations one can conclude that the $\ell_1(Dx)/\ell_2(Dx)$ is not a suitable prior for all the images.

In order to find an explanation for this image dependent behaviour we looked at the following measures: (a) texture (b) total variation and (c) nature of gradient maps, for 55 different images from the database available at [28] and a database of textured images which we collected from the internet. By using Haralick measure [47] for texture, we ruled out the possibility of texture variation being the reason for the image dependent behavior of $R_s(x)$. Haralick measure is calculated in the spatial domain by taking into consideration the statistical nature of texture. Gray tone spatial dependence probability matrices – the co-occurrence matrices – (for the horizontal, vertical and two diagonal directions) are calculated for a given image block and 14 textural features are calculated from these matrices. Few of the features are: angular second moment, sum variance and entropy, difference variance and entropy, etc. We also looked at the total variation of the images. But there was no correlation between the total variation changes and the behaviour of $R_s(x)$ for the different images. Analyzing the horizontal and vertical gradient maps, we observed that those images which have a dense gradient map are those for which the ℓ_1/ℓ_2 norm displays a monotonically rising cost, this is so irrespective of the strength of the edges in the image. Such images we classify as Type 1 images. For images which have a sparse gradient map, behavior of the normalized sparsity measure changes depending on the strength of edges. For images which have strong edges but a sparse gradient map, the ℓ_1/ℓ_2 norm shows a non-monotonic variation with the amount of blur. These images are classified as Type 2a images. We classify images with a sparse gradient map and weak edges as Type 2b – the ℓ_1/ℓ_2 norm decreases monotonically with the amount of blur for these images. The images in Fig. 7.1a, c, e are, respectively, Type 1, Type 2a, and Type 2b images. The horizontal and vertical gradients for these three images are shown in Fig. 7.3.

A minimum value of unity for $R_s(y)$ implies that the minimization process favors the sharp unblurred image as the solution. But as seen from Fig. 7.2, only for a Type 1 image the $R_s(y)$ has unity as the minimum. For Type 2a images, this regularizer performs tolerably well if the PSF size is large. A non-monotonic behavior exists for small blur size. Hence for such an image blurred by a large amount, an acceptable level of deblurring does occur though the minimum is not at the sharp image. For Type 2b images, this regularizer is not suitable since the cost

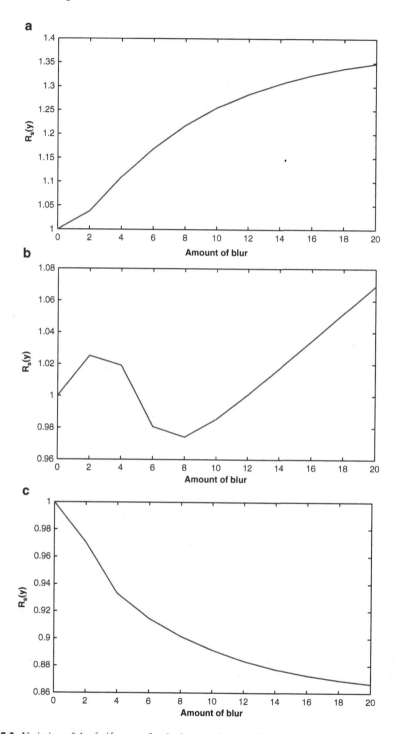

Fig. 7.2 Variation of the ℓ_1/ℓ_2 norm for the images shown in Fig. 7.1

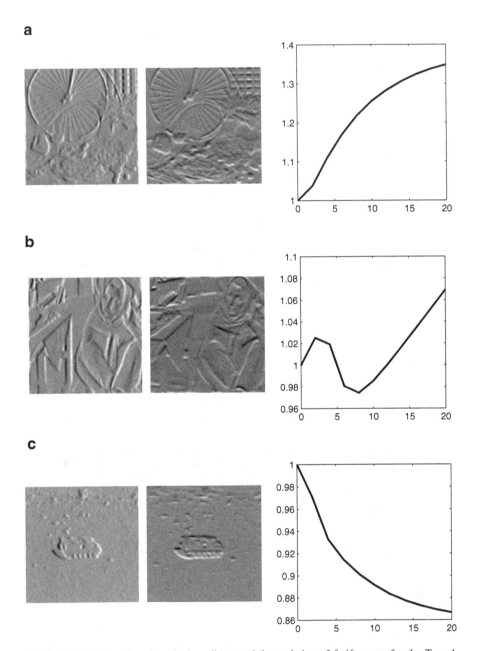

Fig. 7.3 The horizontal and vertical gradients and the variation of ℓ_1/ℓ_2 norm for the Type 1, Type 2a and Type 2b images in Fig. 7.1a, c, e

is monotonically decreasing from the maximum of unity. From Fig. 7.2, it can be observed that $R_s(x)$ is not a good regularizer since its behavior depends on the type of the image.

The intuition behind ℓ_1/ℓ_2 norm is that the ℓ_1 part forces the solution towards a sparse one [32, 76]. The ℓ_2 part makes the ℓ_1/ℓ_2 norm scale invariant. To show that the ℓ_1/ℓ_2 norm does select a sparse solution, we consider a vector $\underline{x} = [x_1, x_2]$ of two elements. The ℓ_1/ℓ_2 norm of this vector is

$$\frac{\ell_1(\underline{x})}{\ell_2(\underline{x})} = \frac{|x_1| + |x_2|}{\sqrt{x_1^2 + x_2^2}}. \tag{7.3}$$

We find the stationary points of the function in Eq. (7.3) in the first quadrant ($x_1 \geq 0, x_2 \geq 0$). The gradient of Eq. (7.3) is

$$\nabla\left(\frac{\ell_1(\underline{x})}{\ell_2(\underline{x})}\right) = \begin{pmatrix} \dfrac{x_1(x_2 - x_1)}{(x_1^2 + x_2^2)^{3/2}} \\ \dfrac{x_1(x_1 - x_2)}{(x_1^2 + x_2^2)^{3/2}} \end{pmatrix}. \tag{7.4}$$

From Eq. (7.4), the stationary points are $x_1 = 0$, $x_2 = 0$, and $x_1 = x_2$. The former two points correspond to the minima and the latter to the maxima. Hence the function attains its minimum either at $\underline{x} = [0, x_2]$ or at $\underline{x} = [x_1, 0]$, which are both sparse in the two dimensional space. Generalizing, it can be argued that the function reaches its minima when the vector is sparse (in this case at the coordinate axes). It may be noted that for a given vector $\underline{x}$, $\ell_2(\underline{x}) \leq \ell_1(\underline{x})$ always leading to a minimum value of unity for the ℓ_1/ℓ_2 norm, and the values less than unity seen in Fig. 7.2 for the Type 2b image is only due to normalization w.r.t. the respective norms of the original image. To check the suitability of $R_s(x)$ as a regularizer, it needs to be checked whether it has a monotonically rising cost as a function of the blur amount.

Claim 7.1. $R_s(x) = \ell_1(Dx)/\ell_2(Dx)$ *is a non-monotonic function of the amount of blur.*

We give a sketch of the proof below.

Proof. Since $R_s(x)$ is the ratio of the ℓ_1 norm to ℓ_2 norm, its monotonicity depends on the relative rate at which the two individual norms decrease with the blur. Let σ be the parameter which indicates the amount of blur. For Gaussian blur, σ is the spread of the Gaussian. Let

$$b(\sigma) \triangleq \ell_1(Dx), \qquad c(\sigma) \triangleq \ell_2(Dx), \tag{7.5}$$

To prove that $R_s(.)$ is monotonically rising, one has to show that

$$\frac{b'(\sigma)}{b(\sigma)} > \frac{c'(\sigma)}{c(\sigma)}, \qquad \forall \sigma, \tag{7.6}$$

where the prime indicates the derivative. We show that Eq. (7.6) fails to hold for $R_s(x)$ for the simple case of a one dimensional signal convolved with a Gaussian impulse response. Let $x(n)$ be the original signal, $h(n)$ be the impulse response of a linear shift invariant system, and $y(n) = x(n) \circledast h(n)$ where $\circledast$ is the convolution operation. For a Gaussian $h(n)$ with spread σ

$$y'(n) = \frac{1}{C} \sum_k \exp(-\frac{k^2}{2\sigma^2}) x(n-k), \tag{7.7}$$

where C is the normalizing constant for the Gaussian impulse response. Since $b(.)$ and $c(.)$ are respectively, the ℓ_1- and ℓ_2-norms, they can be written as

$$b(\sigma) = \sum_n |y'(n)|, \qquad c(\sigma) = \left(\sum_n y'^2(n)\right)^{1/2}. \tag{7.8}$$

Further it can be shown for the Gaussian case, that

$$\frac{b'(\sigma)}{b(\sigma)} = A \frac{\sum_n \mathrm{sgn}(y'(n)) d(n)}{\sum_n |y'(n)|}, \tag{7.9}$$

$$\frac{c'(\sigma)}{c(\sigma)} = A \frac{\sum_n y'(n) d(n)}{\sum_n y'^2(n)}, \tag{7.10}$$

where A is a constant which depends on the normalizing factor C of $h(n)$ and its spread σ, $\mathrm{sgn}(.)$ is the signum function, and $d(n)$ is

$$d(n) = \sum_k k^2 h(k) x(n-k).$$

Using Eqs. (7.9) and (7.10), the inequality in eqn (7.6) becomes

$$\frac{\sum_n \mathrm{sgn}(y'(n)) d(n)}{\sum_n |y'(n)|} > \frac{\sum_n y'(n) d(n)}{\sum_n y'^2(n)}. \tag{7.11}$$

This inequality fails when $y'(n)$ is sparse and has low values which is the case for Type 2 images. Hence $R_s(x)$ is not monotonically increasing and is not an appropriate regularizer.   □

Owing to the disadvantage of image dependent cost of $R_s(x)$, we look at other possibilities and define two new regularizers: one in the derivative domain and the other in the wavelet domain. The details of how we select the two regularizers are given next.

7.1.2 Derivative Domain Regularizer

As seen from the last section, $R_s(x)$ fails for Type 2a and 2b images, since the individual components in the first-order difference $(y'(n))$ has low values. One way to ensure that the inequality in Eq. (7.11) is satisfied is to ensure that the denominator of the R.H.S. of the inequality is sufficiently high. For a Type 2 image, the first-order difference would be quite sparse and taking a second-order difference would lead to increase of component values in the difference signal. This leads to a choice of the ℓ_2 norm of the second derivative as a more appropriate normalization factor, leading to the modified regularizer

$$R_{ds}(x) = \frac{\ell_1(Dx)}{\ell_2(D^2x)}, \qquad (7.12)$$

where D^2x is the second-order difference of the image x. For this norm, the inequality in Eq. (7.11) becomes

$$\frac{\sum_n \text{sgn}(y'(n))d(n)}{\sum_n |y'(n)|} > \frac{\sum_n y''(n)d('n)}{\sum_n y''^2(n)}, \qquad (7.13)$$

where $y''(n)$ is the second order difference of $y(n)$. Since $y''(n)$ has elements with larger absolute value than $y'(n)$, the inequality will be satisfied even for Type 2b images. We verify this experimentally by plotting the variation of $R_{ds}(.)$ (Fig. 7.4). The variation on the ℓ_1 norm of derivative and ℓ_2 norm of double derivative is also plotted in Fig. 7.4 to show that the ℓ_2 norm of the second derivative falls off sufficiently fast enough to keep the ratio of ℓ_1 norm of derivative to ℓ_2 norm of the second derivative a monotonically rising function, which makes it suitable as a regularizer for the image. It may also be noted that for the Type 2b image, the range of variation of the cost is much higher compared to that of the norm proposed in [76] making $R_{ds}(x)$ a more suitable regularizer than $R_s(x)$. In fact this is true for all the three types of images.

Since the Laplacian gives an average of the second order difference along the horizontal (Δ_{xx}) and vertical (Δ_{yy}) directions, it is not a suitable second derivative. In addition to Δ_{xx}, Δ_{yy} we also use the cross terms ($\Delta_{xy} = \Delta_{yx}$) for evaluating the second derivative. From Fig. 7.4 it is seen that behavior of this regularizer is similar for all the three types of images. We give additional results for the three types of images demonstrating the nature of the two regularizers for images with different sparsity levels for the gradient. Figure 7.5 shows that the new regularizer and the regularizer in [76] both show a monotonically non-decreasing cost for Type 1 images. For Type 2a images shown in Fig. 7.6 it is seen that the regularizer in [76] shows an initial decrease in the cost, but later increases as the blur amount increases. As seen from Fig. 7.7 for Type 2b images, the cost of the regularizer in [76] decreases monotonically. The regularizer in Eq. (7.12) has a cost that increases with the amount of blur for both the Type 2 images as seen from Figs. 7.6 and 7.7.

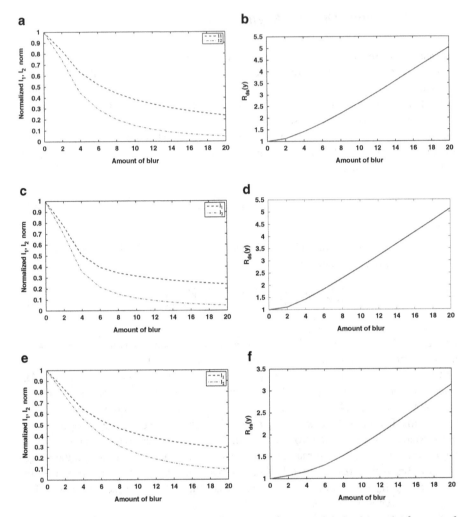

Fig. 7.4 Variation of the ℓ_1, ℓ_2 norms and the ratio of ℓ_1 norm of derivative to the ℓ_2 norm of double derivative. (**a**) and (**b**) Type 1, (**c**) and (**d**) Type 2a, and (**e**) and (**f**) Type 2b images shown in Fig. 7.1

We have also observed that if the condition in Eq. (7.13) is not satisfied for ℓ_2 norm of double derivative, higher order derivatives could be used in the place of double derivative.

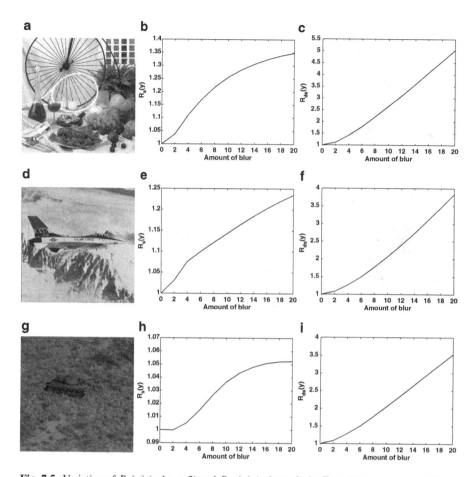

Fig. 7.5 Variation of $R_s(y)$ (column 2) and $R_{ds}(y)$ (column 3) for Type 1 images (column 1)

7.1.3 Wavelet Domain Regularizer

Another sparse domain which could be used is the wavelet domain. While choosing the sparse domain, it should be ensured that the image formation equation (7.14) is valid in the sparse domain too.

$$\underline{y} = K\underline{x} + \underline{n}, \qquad (7.14)$$

If discrete wavelet transform (DWT) representation is used, then Eq. (7.14) is not valid in the transform domain since the convolution matrix and DWT matrix do not commute, i.e.,

$$W\underline{y} \neq KW\underline{x} + \underline{n}, \qquad (7.15)$$

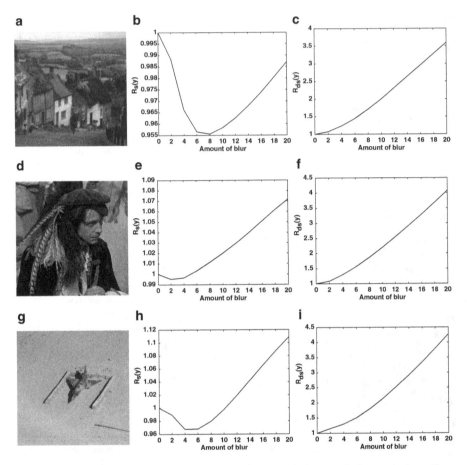

Fig. 7.6 Variation of $R_s(y)$ (column 2) and $R_{ds}(y)$ (column 3) for Type 2a images (column 1)

where W is the DWT matrix. This problem can be overcome by using the stationary wavelet transform (SWT) [138]. SWT is an undecimated version of DWT which involves filtering the rows and columns using high pass and low pass filters. This makes the SWT operation commute with the convolution matrix enabling us to write

$$W_s \underline{y} = K W_s \underline{x} + \underline{n}, \tag{7.16}$$

where $W_s \underline{y}$ is the stationary wavelet transform of $\underline{y}$. The first level SWT of an $M \times N$ image gives four sub-bands similar to the DWT, with the difference that each of the sub-bands is of the same size as the image ($M \times N$). We have used only the level 1 decomposition though it can be extended to higher levels.

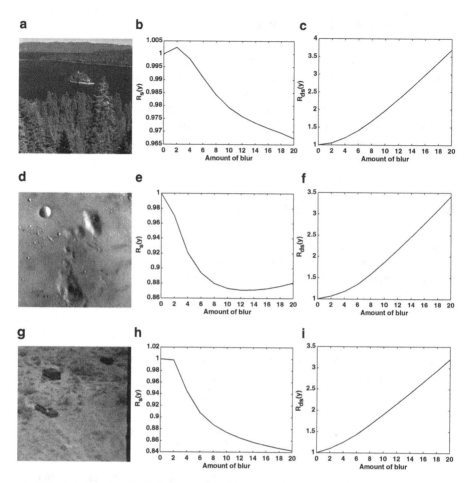

Fig. 7.7 Variation of $R_s(y)$ (column 2) and $R_{ds}(y)$ (column 3) for Type 2b images (column 1)

Out of the four sub-bands, one is dense, generated by two low pass filtering operations and the other three are sparse, generated using LH, HL and HH filtering (L – low pass, H – high pass). For the sparse sub-bands, the ratio of ℓ_1 norm to ℓ_2 norm of the sub-bands is a suitable regularizer for Type 1 images. This norm again has the disadvantage that it does not work for Type 2 images. For the dense sub-band, taking the ℓ_1 norm is same as finding the average of the sub-band, hence ℓ_1/ℓ_2 norm is not a suitable regularizer for the dense frame. Hence we use the quadratic upper bounded TV regularization for the dense sub-band. The final estimate of PSF is taken as the average of estimates with both regularizers.

7.2 Deconvolution Framework

The method adopted for implementing the optimization with the regularizers mentioned in Sect. 7.1 is described here. Without loss of generality, the derivative domain regularizers derived here are used to deconvolve a motion-blurred image and the wavelet domain regularizers are used to deconvolve an image degraded by a Gaussian blur. In both the cases, the PSF is estimated initially and this estimate is used to estimate the image using a non-blind deconvolution procedure so that one can compare the results with those by Krishnan et al. [76].

7.2.1 Cost Function for Derivative Domain Regularizer

Estimation is done in the sparse derivative domain. The image formation model of Eq. (1.9) is converted to the derivative domain by applying the horizontal and vertical difference operators, $\nabla_h = [-1, 1]$ and $\nabla_v = [1, -1]^T$, respectively, to Eq. (1.9) [2,76]. This gives

$$\underline{g} = K\underline{f} + \hat{\underline{n}}, \tag{7.17}$$

where $\underline{g} = [\nabla_h \underline{y}, \nabla_v \underline{y}]^T$ is the gradient of the observed image and $\underline{f} = [\nabla_h \underline{x}, \nabla_v \underline{x}]^T$ is the gradient of the original image. Since K and the derivative are both LSI systems they commute ($\nabla_h K\underline{x} = K\nabla_h \underline{x}$). Now, the cost function becomes

$$C(\underline{f}, \underline{k}) = \parallel K\underline{f} - \underline{g} \parallel_2^2 + \lambda_{\underline{f}} \frac{\parallel \underline{f} \parallel_1}{\parallel [\nabla_h \underline{f}, \nabla_v \underline{f}] \parallel_2} + \lambda_{\underline{k}} \parallel \underline{k} \parallel_1, \tag{7.18}$$

where $\parallel . \parallel_2$ and $\parallel . \parallel_1$, are respectively, the ℓ_2 and ℓ_1 norms, and λ_f and λ_k are the image and the PSF regularization factors. Since motion blurs are typically sparse in nature, the ℓ_1 norm was selected as the PSF regularizer. As discussed in Chap. 1, this cost function is minimized using AM. Intermediate estimates of $\underline{f}$ and $\underline{k}$ are made by alternately keeping each of the variables fixed and estimating the other. The final estimate of $\underline{f}$ is discarded, since the reconstructed image using the Poisson solver showed lot of inconsistencies. This could be due to the ℓ_2 norm of the double-derivative in the denominator which can lead to noise amplification.

7.2.1.1 Updating $\underline{f}$

While updating $\underline{f}, \underline{k}$ is considered to be a constant. This reduces the cost to

$$C(\underline{f}) = \parallel K\underline{f} - \underline{g} \parallel_2^2 + \lambda_{\underline{f}} \frac{\parallel \underline{f} \parallel_1}{\parallel [\nabla_h \underline{f}, \nabla_v \underline{f}] \parallel_2}. \tag{7.19}$$

Presence of the ℓ_2 norm in the denominator of the regularizer makes the cost function non-convex. Hence its value is evaluated at each iteration using the previous estimate of $\underline{f}$. With this approximation, the cost is of the form $C(x) = u(x)+v(x)$, where u is a quadratic function and v is the ℓ_1 norm. This is similar to the scenario described in Sect. 2.6 and ISTA is used to find the minimum of Eq. (7.19). The update mechanism described in Eq. (2.66) is repeated here for convenience.

$$\underline{f}_i = \mathscr{T}_\alpha(\underline{f}_{i-1} - \alpha\nabla C(\underline{f}_{i-1})),$$
$$= \mathscr{T}_\alpha(\underline{f}_{i-1} - \alpha K^T(K\underline{f}_{i-1} - \underline{g})), \tag{7.20}$$

where $\alpha = \dfrac{\| [\nabla_h\underline{f}_{i-1}, \nabla_v\underline{f}_{i-1}] \|_2}{\lambda_{\underline{f}}}$, i counts the iteration, and $\mathscr{T}_\alpha$ is the shrinkage operator

$$\mathscr{T}_\alpha(\underline{f})_i = \max(|\underline{f}_i| - \alpha, 0)\mathrm{sgn}(\underline{f}_i), \tag{7.21}$$

where sgn(.) is the signum function. Two levels of iteration are used to estimate $\underline{f}$, the inner level uses the ISTA step and the outer level is used to update α which changes due to the ℓ_2 norm factor.

7.2.1.2 Updating $\underline{k}$

Here $\underline{f}$ is kept constant, which modifies the cost as

$$C(k) = \| F\underline{k} - g \|_2^2 + \lambda_k \| \underline{k} \|_1 . \tag{7.22}$$

We have used a majorization-minimization approach to solve for k [76,85]. A fast deconvolution procedure using hyper-Laplacian prior [75] is used to estimate x from y and the estimated PSF (k).

7.2.2 Cost Function for Wavelet Domain Regularizer

Without loss of generality we use the Gaussian blur to test the wavelet domain prior. A single level decomposition using the stationary wavelet transform (SWT) is used to get four sub-bands. Applying SWT on the image formation model Eq. (1.9) gives

$$Y_s = KX_s + \hat{N}, \tag{7.23}$$

where $Y_s = [y_{ll}, y_{lh}, y_{hl}, y_{hh}]$; y_{ll} is the sub-band obtained through row- and column-wise low pass filtering, y_{lh} through row-wise low pass and column-wise high pass filtering, y_{hl} through row-wise high pass and column-wise low pass filtering, and y_{hh} through row- and column-wise high pass filtering. X_s is defined similarly for the observed image. Since different regularizers are used for the dense and the sparse sub-bands, the dense sub-band is considered as a separate image and the blind deconvolution is done on it similar to the method in Chap. 3, with quadratic upper-bounded TV as the regularizer. For the sparse sub-bands, an approach similar to the one in the previous subsection is used with a slight modification in the image regularizer. The image regularizer used in this case is the ratio of ℓ_1 norm to ℓ_2 norm of the sparse sub-bands. The PSF regularizer used is the total variation of the blur. The cost function for the dense sub-band is

$$C_d(x_{ll}, k) = \| K\underline{x}_{ll} - \underline{y}_{ll} \|_2^2 + \underline{x}_{ll}^T D^T \Lambda_x D\underline{x}_{ll} + \underline{k}^T D^T \Lambda_k D\underline{k}. \qquad (7.24)$$

The cost for a sparse sub-band is

$$C_d(x_s, k) = \| Kx_s - y_s \|_2^2 + \lambda_x \frac{\| x_s \|_1}{\| x_s \|_2} + \underline{k}^T D^T \Lambda_k D\underline{k}, \qquad (7.25)$$

where $x_s = \{x_{lh}, x_{hl}, x_{hh}\}$, and each component is updated independently using the corresponding image prior. The update equations are similar to that of previous section and the PSF is estimated using conjugate gradient method. The results are given in Sect. 7.3.

7.3 Experimentation

The deconvolution results for the regularizers which we have derived are given in this section. The derivative domain regularizer was tested for a real, motion blurred data given in Fig. 7.8a. The reconstructed image using the regularizer in [76] and the modified regularizer is shown in Fig. 7.8b, c, respectively. The reconstructed kernels are shown alongside with the restored images. For Type 1 image, performance comparable to that of [76] is obtained, as seen from Fig. 7.8. In addition to the regularization factor λ_f, the number of iterations also tend to affect the quality of reconstruction, which leads to the number of parameters to be tuned rather high. The inner and outer iterations were limited to be less than four. The PSF and image regularization factors had to be fixed by trial and error.

Another set of results for motion blurred Type 1 images is given in Fig. 7.9. It is seen from Figs. 7.8 and 7.9 that, as expected, the modified regularizer gives a performance comparable to that of the regularizer in [76] for images which are of Type 1. Another data set which depicts a real camera shake is given in Fig. 7.10, which shows the deblurring due to the modified regularizer.

The results for Type 2b images are given in Figs. 7.11 and 7.12 along with the estimated kernels. The two blurred images, are created synthetically using the kernels shown along with the blurred images. For the image in Fig. 7.11, there is

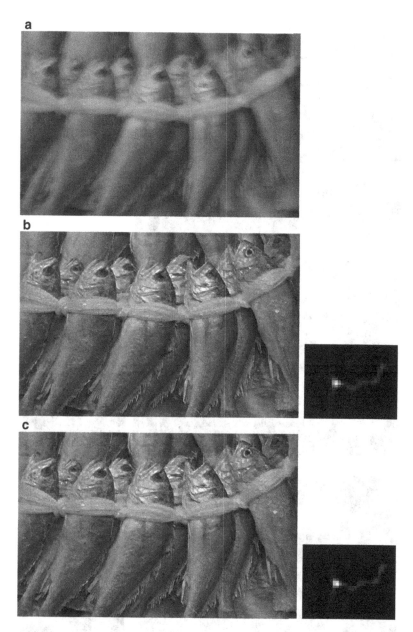

Fig. 7.8 (a) Blurred image. (b) Deblurred image and estimated kernel using regularizer in [76] (c) Deblurred image and estimated kernel using the modified derivative domain regularizer (Image Courtesy: Krishnan et al. [76])

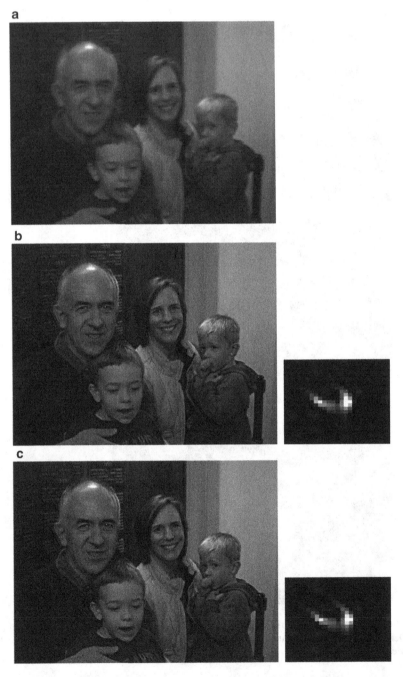

Fig. 7.9 (a) Blurred image. (b) Deblurred image and estimated kernel using regularizer in [76] (c) Deblurred image and estimated kernel using the modified derivative domain regularizer (Image Courtesy: Krishnan et al. [76])

Fig. 7.10 (a) Blurred image. (b) Deblurred image using the modified derivative domain regularizer (Image Courtesy: Abhimitra Meka, IIT Bombay)

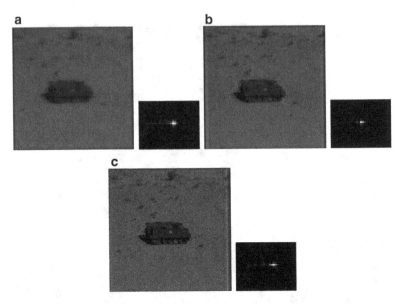

Fig. 7.11 (**a**) Blurred image. (**b**) Deblurred image using regularizer in [76]. (**c**) Deblurred image using the modified derivative domain regularizer. Improvement in PSNR in comparison with [76]: 4 dB (Image courtesy:http://decsai.ugr.es/cvg/CG/base.htm)

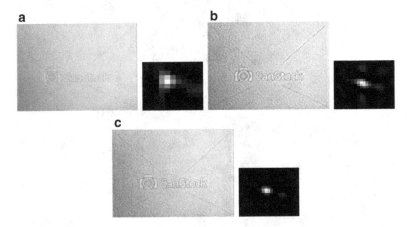

Fig. 7.12 (**a**) Blurred image. (**b**) Deblurred image using regularizer in [76] (**c**) Deblurred image using the modified derivative domain regularizer. Improvement in PSNR in comparison with [76]: 7 dB (Image courtesy: http://www.canstockphoto.com/low-contrast-paper-towel-texture-0957980.html)

a 4 dB improvement over the PSNR compared to the image estimated using the regularizer in [76]. Similarly, there is a 7 dB improvement over [76] for the modified regularizer for the image in Fig. 7.12.

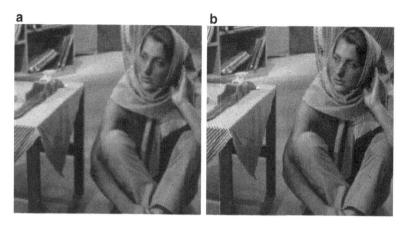

Fig. 7.13 (**a**) Blurred image. (**b**) Deblurred image using the modified wavelet domain regularizer

The wavelet domain regularizer was tested using synthetic data. A noisy and blurred image was generated using a truncated Gaussian blur of size 5 and spread 2, with a noise variance of 5. The result of deblurring using TV regularizer for the dense sub-band and ℓ_1/ℓ_2 norm for the sparse sub-bands is given in Fig. 7.13b, the blurred and noisy image is shown in Fig. 7.13a. Observing Fig. 7.13b, it may be noted that the lines on the table cloth and dress have become visible.

7.4 Discussions

One method of preventing a trivial solution in joint MAP estimation is to use an image regularizer which has a cost that increases with the amount of blur. We observe that the derivative domain ℓ_1/ℓ_2 norm proposed by Krishnan et al., does not exhibit the expected monotonically increasing cost for all types of images. For Type 2b images, this norm is a monotonically decreasing one. We define a derivative domain regularizer: the ratio of ℓ_1 norm of the image derivative to the ℓ_2 norm of second derivative of image. This regularizer has a cost which is monotonically rising for all the three different types of images. We also derive a condition to check the suitability of the derivative domain norms. From the experimental results it is seen that the new regularizer shows comparable results in the case of Type 1 images and an improvement in PSNR for Type 2a and 2b images compared to that obtained using the ℓ_1/ℓ_2 norm of derivative. We also derive a stationary wavelet transform domain regularizer. Since one of the sub-bands generated by the wavelet transform is a dense sub-band, TV regularizer is used for the same and for the sparse sub-bands the ℓ_1/ℓ_2 norm is used.

Chapter 8
Conclusions and Future Research Directions

The findings of our exploration of blind deconvolution algorithms are provided in this chapter. We also give some directions for future research.

8.1 Conclusions

There exists a claim in the recent literature that joint MAP gives only trivial results and hence cannot be used for blind deconvolution. In addition, unlike the deblurring problem where state-of-the-art algorithms have rigorous convergence analysis, the blind deconvolution problem has no such claim. This is due to the highly ill-posed nature of the blind deconvolution problem. Lack of convergence analysis and analyzing the suitability of joint MAP were some of the main motivating factors for the research which form the basis of this monograph.

We explained methods for selecting appropriate PSF regularizers and the regularization factors. In addition an image prior that guarantees a non-trivial solution in joint estimation is also derived. We have also carried out the convergence studies of alternate minimization algorithm for jointly estimating the image and the PSF in blind deconvolution.

We started by proving that choosing an appropriate PSF regularizer will eliminate trivial results during a joint estimation of the image and the PSF. A necessary condition on the PSF regularizer is that it attains its maximum value at the 2D-discrete impulse. This condition alone is not sufficient to prevent the trivial solution and the PSF regularization factor needs to be selected appropriately. We derived an acceptable range for the PSF regularization factor, with an exact lower bound and an approximate upper bound. This provides a way to select a suitable regularization factor without resorting to computationally intensive methods like the generalized cross validation.

© Springer International Publishing Switzerland 2014
S. Chaudhuri et al., *Blind Image Deconvolution: Methods and Convergence*,
DOI 10.1007/978-3-319-10485-0_8

Having proved that joint MAP does give non-trivial solutions, we looked at the convergence of alternate minimization (AM) for blind deconvolution for the smoothness regularizer as well as the quadratic upper-bounded total variation (TV) regularizer. Convergence of AM with TV as a regularizer was carried out in the Fourier domain. A quadratic upper-bounded TV function was used to make the cost quadratic at each iteration and a further approximation was made to make the system a linear shift invariant one. This led to an iterative process in which the regularization factors change with each iteration, reaching a constant value once the fixed point is attained. The resulting image and the PSF spectral magnitudes are related to each other. This occurs due to the approximation used to make the system shift invariant. This part of the analysis also shows that the TV regularization works like an adaptive Wiener filter. We also analyze the error incurred by making the system shift invariant. This analysis shows that high frequency information is lost due to the approximation and that the quadratic upper bounded TV works by adding regularized higher derivatives of image points with low gradient, thus giving rise to a spatially varying regularization.

We also explored spatial domain methods for proving the convergence of alternate minimization algorithm. We extended the geometric proof of convergence of AM, normally used for problems in information theory, which use KL divergence as the cost function to blind deconvolution. We showed that such an approach to convergence analysis of AM for blind deconvolution is possible by demonstrating it for the smoothness prior. We showed that all the points in the image space satisfy the three-point property and derived an expression for the non-negative function needed to define the three- and four-point properties. We also showed that under certain conditions all points in the image space satisfy the four point property. With both properties satisfied the sequence of image and PSF vectors generated by the AM algorithm converge to the point at which the cost function reaches its infimum.

From the two types of convergence analysis, it is inferred that the image estimation part of the AM is more amenable to analysis. This could be due to the fact that the convolution matrix in this case arises from the PSF. This is a low pass filter, leading to a matrix with predictable properties, making analysis easier. This is not so in the step where PSF is estimated, since the convolution matrix arises from the observed image.

An alternative to using appropriate PSF regularizer for preventing trivial solutions is to use an image regularizer which has a cost that increases with the blur amount. We derive a condition using which the suitability of ℓ_1/ℓ_2 norm as a regularizer can be checked for ℓ_1 norm of derivative as the numerator and ℓ_2 norm of higher order derivatives as the denominator. We show that the ℓ_1/ℓ_2 norm of derivative does not exhibit uniform behavior across different types of images. We use the ℓ_1 of derivative to ℓ_2 of double derivative as an alternative that shows a monotonically increasing cost with the blur amount for all types of images. In addition, we define a prior in the wavelet domain, which is another sparse domain. The ℓ_1/ℓ_2 norm proves to be a suitable regularizer provided the order of the derivative of the image on which the ℓ_2 norm operates is selected appropriately. If the image has weak edges then one needs to take the ℓ_2 norm of higher order derivatives of the image.

8.2   Future Research Directions

It was shown in Claim 4.2 that, in order to get a non-trivial solution of the joint MAP estimation, the regularization factor λ_k for the PSF regularizer must be chosen such that

$$\lambda_k > \frac{\lambda_x E(R_x)}{R_{k_{max}}}.$$

This shows that the choice of λ_k is image content dependent. However $E(R_x)$ is not known and we have assumed $E(R_x) \approx E(R_y)$ where y is the observed image. As the blurring becomes more severe $E(R_x) >> E(R_y)$ and such an assumption is no longer valid. Under this assumption $\lambda_{k_{min}}$ will always be under estimated and the joint estimation may lead to a trivial solution. It may be a good exercise to obtain a better estimate of $E(R_x)$ so that the computed lower bound of λ_k is more meaningful.

One of the main themes of this monograph has been the study of convergence properties of alternate minimization (AM) algorithm for blind deconvolution. Quite naturally, the convergence would depend on the choice of prior. We have demonstrated the convergence for quadratic (smoothness) and approximate TV (quadratically upper bounded TV) priors. Although it may not be possible to prove the convergence for any arbitrary prior, it is worth pursuing research on the issue of identifying a class of priors for which convergence of AM methods is guaranteed.

For the analysis in Fourier domain, we resorted to considering reciprocal of the largest gradient of the estimated image as the step size in Eq. (5.46) to prove the convergence of quadratically upper bounded TV. It did allow us to prove the convergence, but our belief is that the rate of convergence would subsequently be much more slower than that in the case of a standard quadratic smoothness regularizer. Although this is not surprising, the open issue remains whether such a choice is overly conservative. Also since the weight changes at each iteration as the image is progressively more and more restored, we expect the weight to gradually reduce, slowing down the convergence as the iterations proceeds. It would be interesting to investigate the issue of rate of convergence and its dependence on the choice of weight.

In Sect. 5.3, while performing an error analysis due to choice of weight as the reciprocal of the largest gradient, we assumed that the $|P| < 1$ where $P = (K^T K + \lambda_{min} D^T D)^{-1} D^T \Lambda_{res} D$, (see Eq. (5.60)). We have not explored the conditions under which the above assumption is satisfied. In case the above assumption is not satisfied, an alternate method of error analysis need to be investigated.

While analyzing the convergence of AM method using three-point and four-point properties, we have assumed the cost function to be the standard quadratic smoothness criterion for both the image and the PSF terms. We have not explored if a TV norm would be amenable to such an analysis. It will be even more interesting if the convergence of sparsity inducing norm is studied in detail.

In this monograph we have also investigated the issue of selecting sparsity as a possible regularizer. We did show through ample illustrations why $\ell_1(Dx)/\ell_2(D^2x)$ is a better regularizer compared to $\ell_1(Dx)/\ell_2(Dx)$. We did give a sketch of our approach in deriving the monotonicity of the ℓ_1/ℓ_2 function with respect to the blur amount. However, it was not a proof by itself. Notwithstanding above it does allow us to explore if the function $\ell_1(Dx)/\ell_2(D^2x)$ is more suitable in case the $\ell_1(Dx)/\ell_2(Dx)$ function is not monotonic for a image with very weak edges. As a matter of fact one may be able to generalize the criterion to $\ell_1(Dx)/\ell_2(D^nx)$ for a suitable n. However, presence of noise in the observation will make such a criterion very fragile. The robustness issues of such a class of ℓ_1/ℓ_2 sparsity inducing norm remains to be studied.

References

1. M. S. C. Almeida and L. B. Almeida. Blind and semi-blind deblurring of natural images. *IEEE Transactions on Image Processing*, 19(1):36–52, Jan 2010.
2. L. Anat, W. Yair, D. Frédo, and T. F. William. Understanding and evaluating blind deconvolution algorithms. In *CVPR*. IEEE, 2009.
3. G. L. Anderson and A. N. Netravali. Image restoration based on a subjective criterion. *IEEE Transactions on Systems, Man, and Cybernetics*, SMC-6(12):845–853, Dec 1976.
4. H. C. Andrews and B. R. Hunt. *Digital Image Restoration*. Prentice Hall, 1977.
5. G. R. Ayers and J. C. Dainty. Iterative blind deconvolution method and its applications. *Optics Letters*, 13(7):547–549, 1988.
6. D. Babacan, R. Molina, and A. K. Katsaggelos. Variational Bayesian blind deconvolution using a total variation prior. *IEEE Transactions on Image Processing*, 18(1):12–26, 2009.
7. L. Bar, N. Sochen, and N. Kiryati. Semi-blind image restoration via Mumford-Shah regularization. *IEEE Transactions on Image Processing*, 15(2):483–493, 2006.
8. A. Beck and M. Teboulle. Fast gradient-based algorithms for constrained total variation image denoising and deblurring problems. *IEEE Transactions on Image Processing*, 18(11):2419–2434, 2009.
9. M. Bertero and P. Boccacci. *Introduction to Inverse Problems in Imaging*. IOP Publishing Ltd., 1998.
10. J. Besag. Spatial interaction and the statistical analysis of lattice systems. *Journal of the Royal Statistical Society. Series B*, 36:192–236, 1974.
11. J. Biemond, F. G. van der Putten, and J. W. Woods. Identification and restoration of images with symmetric noncausal blurs. *IEEE Transaction on Circuits and Systems*, 35(4):385–393, 1988.
12. J. Brodie, I. Daubechies, C. De Mol, D. Giannone, and I. Loris. Sparse and stable Markowitz portfolios. *Proceedings of the National Academy of Sciences of the USA*, 106(30):12267–12272, 2009.
13. M. Cannon. Blind deconvolution of spatially invariant image blurs with phase. *IEEE Transaction on Acoustics, Speech, and Signal Processing*, ASSP-24(1):58–63, 1976.
14. A. S. Carasso. Direst blind deconvolution. *SIAM Journal of Applied Mathematics*, 61(6):1980–2007, 2001.
15. A. Chambolle and P. L. Lions. Image recovery via total variation minimization and related problems. *Numerical Mathematics*, 76:167–188, 1997.
16. T. F. Chan and C. K. Wong. Total variation blind deconvolution. *IEEE Transactions on Image Processing*, 7(3):370–375, 1998.

© Springer International Publishing Switzerland 2014
S. Chaudhuri et al., *Blind Image Deconvolution: Methods and Convergence*,
DOI 10.1007/978-3-319-10485-0

17. T. F. Chan and C. K. Wong. Convergence of the alternating minimization algorithm for blind deconvolution. *Linear Algebra and its Applications*, 316:259–285, 2000.

18. M. M. Chang, A. M. Teklap, and A. T. Erdem. Blur identification using the bispectrum. *IEEE Transaction on Signal Processing*, 39(10):2323–2325, 1991.

19. L. B. Charles. Alternating minimization and alternating projection algorithms: A tutorial. Technical report, University of Massachusetts Lowell, 2011. http://faculty.uml.edu/cbyrne/misc.htm.

20. S. Chaudhuri and A. N. Rajagopalan. *Depth from Defocus – a Real Aperture Imaging Approach*. Springer, 1999.

21. L. Chen and K. Yap. A soft double regularization approach to parametric blind image deconvolution. *IEEE Transactions on Image Processing*, 14(5):624–633, 2005.

22. S. S. Chen, D. L. Donoho, and M. A. Saunders. Atomic decomposition by basis pursuit. *SIAM Journal on Scientific Computing*, 20:33–61, 1998.

23. B. A. Chipman and B. D. Jeffs. Blind multiframe point source image restoration using map estimation. In *Signals, Systems, and Computers, 1999. Conference Record of the Thirty-Third Asilomar Conference on*, volume 2, pages 1267–1271 vol.2, Oct.

24. S. Cho and S. Lee. Fast motion deblurring. *ACM Transactions on Graphics (SIGGRAPH ASIA 2009)*, 28(5):article no. 145, 2009.

25. T. S. Cho, S. Paris, B. K. P. Horn, and W. T. Freeman. Blur kernel estimation using the radon transform. In *CVPR 2008*, 2008.

26. E. K. P. Chong and S. H. Zak. *An Introduction to Optimization, 2nd Edition*. Wiley-Interscience, 2001.

27. I. Csiszar and G. Tusnady. Information geometry and alternating minimization procedures. *Statistics and Decisions Supp. 1*, pages 205–237, 1984.

28. Online Database. Test images. http://decsai.ugr.es/cvg/CG/base.htm.

29. I. Daubechies, M. Defrise, and C. De Mol. An iterative thresholding algorithm for linear inverse problems with a sparsity constraint. *Communications on Pure and Applied Mathematics*, 57(11):1413–1457, 2004.

30. B. L. K. Davey, R. G. Lane, and R. H. T. Bates. Blind deconvolution of noisy complex-valued image. *Optics Communications*, 69:353–356, 1989.

31. P. J. Davis. *Circulant Matrices*. John Wiley & Sons, 1979.

32. M. Elad. *Sparse and Redundant Representations – From Theory to Applications in Signal and Image Processing*. Springer, 2010.

33. H. W. Engl, M. Hanke, and A. Neubauer. *Regularization of Inverse Problems*. Kluwer Academic Publishers, 2000.

34. M. Feder. Statistical signal processing using a class of iterative estimation algorithms. Technical report, Massachusetts Institute of Technology, 1987.

35. W. Feller. *An Introduction to Probability Theory and Its Applications*. Wiley, 1968.

36. R. Fergus, B. Singh, A. Hertzmann, S. T. Roweis, and W.T. Freeman. Removing camera shake from a single photograph. *ACM Transactions on Graphics, SIGGRAPH 2006 Conference Proceedings, Boston, MA*, 25:787–794, 2006.

37. D. J. Field. What is the goal of sensory coding? *Neural Comput.*, 6(4):559–601, July 1994.

38. M. A. T. Figueiredo and J. M. Bioucas-Dias. Restoration of Poissonian images using alternating direction optimization. *IEEE Transactions on Image Processing*, 19(12):3133–3145, 2010.

39. M. A. T. Figueiredo, J. M. Bioucas-Dias, and R. D. Nowak. Majorization-minimization algorithms for wavelet-based image restoration. *IEEE Transactions on Image Processing*, 16(12):2980–2991, 2007.

40. M. A. T. Figueiredo, J. B. Dias, J. P. Oliveira, and R.D. Nowak. On total variation denoising: A new majorization-minimization algorithm and an experimental comparison with wavelet denoising. In *ICIP*, 2006.

41. N. P. Galatsanos and A. K. Katsaggelos. Methods for choosing the regularization parameter and estimating the noise variance in image restoration and their relation. *IEEE Transactions on Image Processing*, 1(3):322–336, 1992.

42. D. C. Ghiglia, L. A. Romero, and G. A. Mastin. Systematic approach to two-dimensional blind deconvolution by zero-sheet separation. *Journal of Optical Society of America A*, 10(5):1024–1036, 1993.

43. G. B. Giannakis and R. W. Heath. Blind identification of multichannel FIR blurs and perfect image restoration. *IEEE Transactions on Image Processing*, 9(11):1877–1896, Nov.

44. J. W. Goodman. Some fundamental properties of speckle. *Journal of Optical Society of America*, 66(11):1145–1150, Nov 1976.

45. D. Graupe, D. J. Krause, and J. B. Moore. Identification of autoregressive moving average parameters of time series. *IEEE Transaction on Automatic Control*, AC-20:104–106, 1975.

46. C. W. Groetsch. *The Theory of Tikhonov Regularization for Fredholm Equations of the First Kind*. Pitman Advanced Publishing Program, 1984.

47. R. M. Haralick, K. Shanmugam, and I. Dinstein. Textural features for image classification. *Systems, Man and Cybernetics, IEEE Transactions on*, SMC-3(6):610–621, November 1973.

48. S. S. Haykin. *Unsupervised Adaptive Filtering: Blind source separation*. Wiley-Interscience publication. Wiley, 2000.

49. T. A. Hearn and L. Reichel. Extensions of the Justen – Ramlau blind deconvolution method. *Advances in Computational Mathematics*, pages 1–27, 2013.

50. W. E. Heinz and K. Philipp. Nonlinear inverse problems: Theoretical aspects and some industrial applications. Tutorial, UCLA, 2003. Inverse Problems: Computational Methods and Emerging Applications Tutorials.

51. Y. M. Huang, M. Ng, and Y. W. Wen. A fast total variation minimization method for image restoration. *Multiscale Modeling & Simulation*, 7(2):774–795, 2008.

52. P. Hung-Ta and A. C. Bovik. On eigenstructure-based direct multichannel blind image restoration. *IEEE Transactions on Image Processing*, 10(10):1434–1446, Oct.

53. B. Hunt. A theorem on the difficulty of numerical deconvolution. *Audio and Electroacoustics, IEEE Transactions on*, 20(1):94–95, mar 1972.

54. B. R. Hunt. The application of constrained least squares estimation to image restoration by digital computer. *IEEE Transactions on Computers*, C-22(9):805–812, 1973.

55. N. Hurley and R. Scott. Comparing measures of sparsity. *IEEE Transactions on Information Theory*, 55(10):4723–4741, Oct.

56. A. Hyvärinen, J. Karhunen, and E. Oja. *Independent Component Analysis*. Adaptive and Learning Systems for Signal Processing, Communications and Control Series. Wiley, 2004.

57. Cannon INC. What is optical image stabilizer?, 2006. http://www.canon.com/bctv/faq/optis.html.

58. A. K. Jain. *Fundamentals of Digital Image Processing*. Prentice-Hall, Inc., Upper Saddle River, NJ, USA, 1989.

59. K. Jeongtae and J. Soohyun. High order statistics based blind deconvolution of bi-level images with unknown intensity values. *Optics Express*, 18:12872–12889, 2010.

60. J. Jia. Single image motion deblurring using transparency. In *2007 IEEE Computer Society Conference on Computer Vision and Pattern Recognition (CVPR 2007), 18–23 June 2007, Minneapolis, Minnesota, USA*. IEEE Computer Society, 2007.

61. N. Joshi, R. Szeliski, and D. J. Kriegman. PSF estimation using sharp edge prediction. In *CVPR*, 2008.

62. L. Justen and R. Ramlau. A non-iterative regularization approach to blind deconvolution. *Inverse Problems*, 22:771–800, 2006.

63. M. G. Kang and A. K. Katsaggelos. Simultaneous iterative image restoration and evaluation of the regularization parameter. *IEEE Transactions on Signal Processing*, 40(9):2329–2334, 1992.

64. R. Kashyap. Inconsistency of the AIC rule for estimating the order of autoregressive models. *IEEE Transactions on Automatic Control*, 25(5):996–998, 1980.

65. A. K. Kastaggelos. Iterative image restoration algorithms. *Optical Engineering*, 28(7):735–748, 1989.

66. A. K. Kastaggelos, J. Biemond, R. W. Schafer, and Mersereau R. M. A regularized iterative image restoration algorithm. *IEEE Transactions on Signal Processing*, 39(4):914–929, 1991.

67. A. K. Kastaggelos and K. T. Lay. Simultaneous identification and restoration of images using maximum likelihood estimation. In *Control and Applications, 1989. Proceedings. ICCON'89. IEEE International Conference on*, pages 236–240, 1989.

68. A. K. Kastaggelos and K.T. Lay. Maximum likelihood blur identification and image restoration using the EM algorithm. *IEEE Transactions on Signal Processing*, 39(3):729–733, 1991.

69. A. K. Katsaggelos, J. Biemond, R. M. Mersereau, and R. W. Schafer. A general formulation of constrained iterative restoration algorithms. In *Acoustics, Speech, and Signal Processing, IEEE International Conference on ICASSP '85.*, volume 10, pages 700–703, 1985.

70. A. K. Katsaggelos, J. Biemond, R. M. Mersereau, and R. W. Schafer. Nonstationary iterative image restoration. In *Acoustics, Speech, and Signal Processing, IEEE International Conference on ICASSP '85.*, volume 10, pages 696–699, 1985.

71. T. Kenig, Z. Kam, and A. Feuer. Blind image deconvolution using machine learning for three-dimensional microscopy. *IEEE Transactions on Pattern Analysis and Machine Intelligence*, 32(12):2191–2204, Dec.

72. F. Krahmer, Y. Lin, B McAdoo, K. Ott, D. Widemannk, and B. Wohlberg. Blind image deconvolution: motion blur estimation. Technical report, Institute of Mathematics and its Applications, University of Minnesota, 2006.

73. R. Kress. *Linear Integral Equations*. Prentice Hall, 1999.

74. E. Kreyszig. *Introductory Functional Analysis with Applications*. Wiley-India, 2006.

75. D. Krishnan and R. Fergus. Fast image deconvolution using hyper-Laplacian priors. In Y. Bengio, D. Schuurmans, J. Lafferty, C. K. I. Williams, and A. Culotta, editors, *Advances in Neural Information Processing Systems 22*, pages 1033–1041. 2009.

76. D. Krishnan, T. Tay, and R. Fergus. Blind deconvolution using a normalized sparsity measure. In *CVPR*, pages 233–240. IEEE, 2011.

77. S. Kullback. *Information theory and statistics*. Dover, 1959.

78. D. Kundur and D. Hatzinakos. A novel blind deconvolution scheme for image restoration using recursive filtering. *IEEE Transactions on Signal Processing*, 46(2):375–390, 1998.

79. R. L. Lagendeijk, J. Biemond, and D. E. Boekee. Regularized iterative image restoration with ringing reduction. *IEEE Transactions on Acoustics Speech and Signal Processing*, 36(12):1874–1888, 1988.

80. R. L. Lagendijk, J. Biemond, and D. E. Boekee. Identification and restoration of noisy blurred images using the expectation-maximization algorithm. *IEEE Transactions on Acoustics, Speech, and Signal Processing*, 38(7), 1990.

81. R. L. Lagendijk, A. K. Katsaggelos, and J. Biemond. Iterative identification and restoration of images. In *Proceedings IEEE International Conference on Acoustics, Speech and Signal Processing*, pages 992–995, 1988.

82. R. G. Lane. Blind deconvolution of speckle images. *Journal of Optical Society of America A*, 9(9):1508–1514, 1992.

83. R. G. Lane and R. H. T. Bates. Automatic multidimensional deconvolution. *Journal of Optical Society of America A*, 4(1):180–188, 1987.

84. R. G. Lane, W. R. Fright, and R.H.T. Bates. Direct phase retrieval. *IEEE Transaction on Acoustics, Speech, and Signal Processing*, ASSP-35(4):520–526, 1987.

85. A. Levin. Blind motion deblurring using image statistics. In B. Schölkopf, J. Platt, and T. Hoffman, editors, *Advances in Neural Information Processing Systems 19*, pages 841–848. MIT Press, Cambridge, MA, 2007.

86. A. Levin, R. Fergus, F. Durand, and W. T. Freeman. Image and depth from a conventional camera with a coded aperture. *ACM Transactions on Graphics*, 26(3), 2007.

87. A. Levin and Y. Weiss. User assisted separation of reflections from a single image using a sparsity prior. *IEEE Transactions on Pattern Analysis and Machine Intelligence*, 29(9):1647–1654, 2007.

88. A. Levin, Y. Weiss, F. Durand, and W. T. Freeman. Understanding and evaluating blind deconvolution algorithms. In *CVPR*. IEEE, 2009.

89. A. Levin, Y. Weiss, F. Durand, and W. T. Freeman. Efficient marginal likelihood optimization in blind deconvolution. In *CVPR*. IEEE, 2011.

90. S. Z. Li. *Markov Random Field Modeling in Image Analysis*. Springer Publishing Company, Incorporated, 3rd edition, 2009.

91. H. Liao and M. K. Ng. Blind deconvolution using generalized cross-validation approach to regularization parameter estimation. *IEEE Transactions on Image Processing*, 20(3):670–680, 2011.

92. A. C. Likas and N. P. Galatsanos. A variational approach for Bayesian blind image deconvolution. *IEEE Transactions on Signal Processing*, 52(8):2222–2233, 2004.

93. B. V. Limaye. *Functional Analysis*. New Age International Publishers, 2 edition, 2012.

94. S. G. Mallat. A theory for multiresolution signal decomposition: The wavelet representation. *IEEE Transactions on Pattern Analysis and Machince Intelligence*, 11(7):674–693, July 1989.

95. M. Manuel and F. Paolo. Single image blind deconvolution with higher-order texture statistics. In *Video Processing and Computational Video*, pages 124–151, 2010.

96. K. May, T. Stathaki, and A. K. Katsaggelos. Blind image restoration using local bound constraints. In *Acoustics, Speech, and Signal Processing, IEEE International Conference on ICASSP '98.*, volume 5, pages 2929–2932, 1998.

97. B. C. McCallum. Blind deconvolution by simulated annealing. *Optics Communications*, 75(2):101–105, 1990.

98. J. M. Mendel. Tutorial on higher-order statistics (spectra) in signal processing and system theory: theoretical results and some applications. *Proceedings of the IEEE*, 79(3):278–305, 1991.

99. F. Meng and G. Zhao. On second-order properties of the Moreau-Yosida regularization for constrained nonsmooth convex programs. *Numerical Functional Analysis and Optimization*, 25:515–529, 2004.

100. R. Mifflin, L. Qi, and D. Sun. Properties of the moreau-yosida regularization of a piecewise c^2 convex function. *Mathematical programming*, 84(2):269–281, 1999.

101. K. Miller. Least squares methods for ill-posed problems with a prescribed bound. *SIAM Journal of Mathematical Analysis*, 1(1):52–74, 1970.

102. J. W. Miskin and David J. C. MacKay. Ensemble learning for blind image separation and deconvolution. In Girolami M, editor, *Advances in Independent Component Analysis*. Springer-Verlag Scientific Publishers, 2000.

103. N. Miura and N. Baba. Extended-object reconstruction with sequential use of the iterative blind deconvolution method. *Optics Communication*, 89:375–379, 1992.

104. M. Moghaddam and M. Jamzad. Motion blur identification in noisy images using fuzzy sets. In *Proceedings of the Fifth IEEE International Symposium on Signal Processing and Information Technology*, 2005.

105. R. Molina, A. K. Katsaggelos, J. Abad, and J. Mateos. A Bayesian approach to blind deconvolution based on Dirichlet distributions. In *Acoustics, Speech, and Signal Processing, 1997. ICASSP-97., 1997 IEEE International Conference on*, volume 4, pages 2809–2812 vol.4, 1997.

106. R. Molina, J. Mateos, and A. K. Katsaggelos. Blind deconvolution using a variational approach to parameter, image, and blur estimation. *IEEE Transactions on Image Processing*, 15(12):3715–3727, December 2006.

107. J. H. Money and S. H. Kang. Total variation minimizing blind deconvolution with shock filter reference. *Image and Vision Computing*, 26(2):302–314, 2008.

108. V. A. Morozov. Linear and nonlinear ill-posed problems. *Matematicheskii Analiz*, 11:129–178, 1973.

109. D. Mumford and J. Shah. Optimal approximations by piecewise smooth functions and associated variational problems. *Communications on Pure and Applied Mathematics*, 42:577–684, 1989.

110. M. T. Nair. *Linear Operator Equations*. World Scientific, 2009.

111. S. K. Nayar and M. Ben-Ezra. Motion-based motion deblurring. *IEEE Transactions on Pattern Analysis and Machine Intelligence*, 26(6):689–698, June.

112. R. Neelamani, H. Choi, and R. G. Baraniuk. Wavelet-domain regularized deconvolution for ill-conditioned systems. In *ICIP*, pages 204–208, 1999.

113. M. K. Ng, R. J. Plemmons, and Q. Sanzheng. Regularization of rif blind image deconvolution. *IEEE Transactions on Image Processing*, 9(6):1130–1134, 2000.

114. C. L. Nikias and J. M. Mendel. Signal processing with higher-order spectra. *IEEE Signal Processing Magazine*, 10(3):10–37, 1993.

115. J. P. Oliveira, J. M. Bioucas-Dias, and M. A. T. Figueiredo. Adaptive total variation image deblurring: A majorization-minimization approach. *Signal Processing*, 89(9):1683–1693, 2009.

116. J. P. A. Oliveira, M. T. Figueiredo, and J. M. Bioucas-Dias. Blind estimation of motion blur parameters for image deconvolution. In *Iberian Conference on Pattern Recognition and Image Analysis*, 2007.

117. Online. ℓ_1 minimization. http://holycrapscience.tumblr.com/post/1241604136/holy-crap-l1-minimization.

118. S. Osher and L. I. Rudin. Feature-oriented image enhancement using shock filters. *SIAM Journal of Numerical Analysis*, 27(4):919–940, 1990.

119. A. Ostaszewski. Fréchet derivative. http://www.maths.lse.ac.uk/Courses/MA409/Notes-Part2.pdf.

120. Pentland A. P. A new sense for depth of field. *IEEE Transactions on Pattern Analysis and Machine Intelligence*, 9(4):532–531, July 1987.

121. G. Panci, P. Campisi, S. Colonnese, and G. Scarano. Multichannel blind image deconvolution using the Bussgang algorithm: Spatial and multiresolution approaches. *IEEE Transactions on Image Processing*, 12(11):1324–1337, 2003.

122. K. L. Patricia. Some recent developments and open problems in solution methods for mathematical inverse problems. In *XXiV National Congress of Applied and Computational Mathematics*, September 2001.

123. K.E. Patrizio Campisi and K. Egiazarian. *Blind Image Deconvolution: Theory and Applications*. CRC Press INC, 2007.

124. D. L. Phillips. A technique for the numerical solution of certain integral equations of the first kind. *Journal of the ACM*, 9(1):84–97, January 1962.

125. S. U. Pillai and B. Liang. Blind image deconvolution using a robust gcd approach. *IEEE Transactions on Image Processing*, 8(2):295–301, Feb.

126. V. Pohl and H. Boche. *Advanced Topics in System and Signal Theory: A Mathematical Approach*. Springer, 2010.

127. Fienup J. R. Phase retrieval algorithms: a comparison. *Applied Optics*, 21(15):2758–2769, 1982.

128. Hunt B. R. A theorem on the difficulty of numerical deconvolulion. *IEEE Transactions on Audio and Electroacoustics*, AU20:94–95, 1972.

129. Hunt B. R. Digital image processing. *Proceedings of the IEEE*, 63(4):693–710, April 1975.

130. A. Rav-Acha and S. Peleg. Two motion-blurred images are better than one. *Pattern Recognition Letters*, 26(3):311–317, February 2005.

131. S. J. Reeves and R. M. Mersereau. Identification of image blur parameters by the method of generalized cross validation. In *IEEE International Symposium on Circuits and Systems 1990*, May 1990.

132. S. J. Reeves and R. M. Mersereau. Optimal estimation of the regularization parameter and stabilizing functional for regularized image restoration. *Optical Engineering*, 29(5):446–454, 1990.

133. S. J. Reeves and R. M. Mersereau. Blur identification by the method of of generalized cross-validation. *IEEE Transaction on Image Processing*, 1(3):301–311, 1992.

134. T. R. Rockafellar. *Convex Analysis*. Princeton University Press, 1972.

135. P. Rodríguez and B. Wohlberg. Efficient minimization method for a generalized total variation functional. *IEEE Transactions on Image Processing*, 18(2):322–332, 2009.

136. S. Roth and M. J. Black. Fields of experts. *International Journal of Computer Vision*, 82(2):205–229, 2009.

137. L. I. Rudin. Images, numerical analysis of singularities and shock filters. Thesis, California Institute of Technology, 1987.

138. R. von Sachs, G. P. Nason, and G. Kroisandt. Spectral representation and estimation for locally stationary wavelet processes. In Serge Dubuc and Gilles Deslauriers, editors, *Spline functions and the theory of wavelets*. CRM Proceesings & Lecture Notes, New York, 1999.

139. A. E. Savakis and H. J. Trussell. Blur identification by residual spectrum matching. *IEEE Transaction on Image Processing*, 2(2):141–151, 1993.

140. R. W. Schafer, R. M. Mersereau, and M. A. Richards. Constrained iterative restoration algorithms. *Proceedings of the IEEE*, 69(4):432–451, 1981.

141. F. C. Schweppe, editor. *Uncertain Dynamic Systems*. Prentice-Hall, 1973.

142. A. Seghouane. A kullback-leibler divergence approach to blind image restoration. *IEEE Transactions on Image Processing*, 20(7):2078–2083, 2011.

143. A.-K. Seghouane and M. Hanif. A kullback-leibler divergence approach for wavelet-based blind image deconvolution. In *Machine Learning for Signal Processing (MLSP), 2012 IEEE International Workshop on*, pages 1–5, Sept.

144. Q. Shan, J. Jia, and A. Agarwala. High-quality motion deblurring from a single image. *ACM Transactions on Graphics (SIGGRAPH)*, 2008.

145. S. Shekhar and H. Xiong, editors. *Encyclopedia of GIS*. Springer, 2008.

146. W. Smith. *Modern Optical Engineering*. McGraw-Hill Professional, 4 edition, 2007.

147. M. M. Sondhi. Image restoration: The removal of spatially invariant degradations. *Proceedings of the IEEE*, 60:842–857, 1972.

148. F. Sroubek, G. Cristobal, and J. Flusser. A unified approach to super-resolution and multichannel blind deconvolution. *IEEE Transactions on Image Processing*, 16(9):2322–2332, Sept.

149. F. Sroubek and J. Flusser. Multichannel blind iterative image restoration. *IEEE Transactions on Image Processing*, 12(9):1094–1106, 2003.

150. F. Sroubek and J. Flusser. Multichannel blind deconvolution of spatially misaligned images. *IEEE Transactions on Image Processing*, 14(7):874–883, July.

151. F. Sroubek and P. Milanfar. Robust multichannel blind deconvolution via fast alternating minimization. *IEEE Transactions on Image Processing*, 21(4):1687–1700, April.

152. I. S. Stefanescu. On the phase retrieval problem in two dimensions. *Journal of Mathematical Physics*, 26(9):2141–2160, 1985.

153. T. G. Stockham, T. M. Cannon, and R. B. Ingebretsen. Blind deconvolution through digital signal processing. *Proceedings of the IEEE*, 63(4):678–693, 1975.

154. A. M. Tekalp, H. Kaufman, and J. W. Woods. Identification of image and blur parameters for the restoration of noncausal blurs. *IEEE Transactions on Acoustics, Speech, and Signal Processing*, ASSP-34(4), 1996.

155. A. M. Thompson, J. C. Brown, J. W. Kay, and Titterington D. M. A study of methods for choosing the smoothing parameter in image restoration by regularization. *IEEE Transactions on Pattern Analysis and Machine Intelligence*, 13(4):326–339, 1991.

156. R. Tibshirani. Regression shrinkage and selection via the LASSO. *Journal of the Royal Statistical Society, Series B*, 58(1):267–288, 1996.

157. A. N. Tikhonov. Solution of incorrectly formulated problems and the regularization method. *Soviet Mathematics*, 4:1035–1038, 1963.

158. F. Tsumuraya, N. Miura, and N. Baba. Iterative blind deconvolution method using Lucy's algorithm. *Astronomy and Astrophysics*, 282:699–708, 1994.

159. D. G. Tzikas, A. C. Likas, and N. P. Galatsanos. Variational Bayesian sparse kernel-based blind image deconvolution with student's-t priors. *IEEE Transactions on Image Processing*, 18(4):753–764, 2009.

160. R. A. Udantha, P. P. Athina, and J. M. Reid. Higher order spectra based deconvolution of ultrasound images. *IEEE Transactions on Ultrasonics, Ferroelectrics and Frequency Control*, 42(6):1064–1075, 1995.

161. C. Vogel and M. Oman. Iterative methods for total variation denoising. *SIAM Journal of Scientific Computing*, 17:227–238, 1996.

162. C. Vonesch and M. Unser. A fast multilevel algorithm for wavelet-regularized image restoration. *IEEE Transactions on Image Processing*, 18(3):509–523, 2009.

163. C. Wang, L. Sun, Z. Chen, J. Zhang, and S. Yang. Multi-scale blind motion deblurring using local minimum. *Inverse Problems*, 26(1):015003, 2010.

164. Y. Wang, J. Yang, W. Yin, and Y. Zhang. A new alternating minimization algorithm for total variation image reconstruction. *SIAM Journal of Imaging Science*, 1(3):248–272, August 2008.

165. Y. Weiss and W. T. Freeman. What makes a good model of natural images? In *CVPR*, 2007.

166. N. Wiener. *Extrapolation, Interpolation, and Smoothing of Stationary Time Series*. The MIT Press, 1964.

167. J. W. Woods and V. K. Ingle. Kalman filtering in two dimensions: Further results. *IEEE Transaction on Acoustics, Speech, and Signal Processing*, ASSP-29(2):188–197, 1981.

168. L. Xinqiao and A. El Gamal. Simultaneous image formation and motion blur restoration via multiple capture. In *Acoustics, Speech, and Signal Processing, 2001. Proceedings. (ICASSP '01). 2001 IEEE International Conference on*, volume 3, pages 1841–1844 vol.3, 2001.

169. Y. Yang, N.P. Galatsanos, and H. Stark. Projection-based blind deconvolution. *Journal of Optical Society of America A*, 11(9):2401–2409, 1994.

170. K.-H. Yap and L. Guan. A computational reinforced learning scheme to blind image deconvolution. *IEEE Transactions on Evolutionary Computation*, 6(1):2–15, Feb.

171. Y. You and M. Kaveh. A regularization approach to joint blur identification and image restoration. *IEEE Transactions on Image Processing*, 5(3):416–428, 1996.

172. Y. L. You and M. Kaveh. A regularization approach to blind restoration of images degraded by shift-variant blurs. In *Acoustics, Speech, and Signal Processing, IEEE International Conference on ICASSP '95.*, volume IV, page 2607–2610, 1995.

173. Y. L. You and M. Kaveh. Anisotropic blind image restoration. In *Image Processing, 1997. ICIP-97., 1997 IEEE International Conference on*, 1996.

174. Y. L. You and M. Kaveh. Ringing reduction in image restoration by orientation-selective regularization. *IEEE Signal Processing Letters*, 3:29–31, 1996.

175. Y. L. You and M. Kaveh. Blind image restoration by anisotropic regularization. *IEEE Transactions on Image Processing*, 8(3):396–407, 1999.

176. Y. L. You and M. Kaveh. Blind image restoration by anisotropic regularization. *IEEE Transactions on Image Processing*, 8:396–407, 1999.

177. D. C. Youla and H. Webb. Image reconstruction by the method of convex projections. *IEEE Transactions on Medical Imaging*, MI-1(2):81–94, October 1982.

178. L. Yuan, J. Sun, L. Quan, and H.-Y. Shum. Image deblurring with blurred/noisy image pairs. *ACM Transactions on Graphics*, 26(3), July 2007.

179. L. Yuan, J. Sun, L. Quan, and H.-Y. Shum. Progressive inter-scale and intra-scale non-blind image deconvolution. *ACM Transactions on Graphics*, 27:74:1-74:10, August 2008.

180. Adami K. Z. Variational methods in Bayesian deconvolution. In *PHYSTAT 2003, SLAC*, pages 143–147, 2003.

181. J. Zhang. The mean field theory in EM procedures for blind Markov random field image restoration. *IEEE Transactions on Image Processing*, 2(1):27–40, Jan.

182. W. Zhao and A. Pope. Image restoration under significant additive noise. *IEEE Signal Processing Letters*, 14(6):401–404, 2007.

Index

Printed in the United States
By Bookmasters